T0224007

Development of Science Teachers' TPACK

Ying-Shao Hsu

Editor

Development of Science Teachers' TPACK

East Asian Practices

 Springer

Editor
Ying-Shao Hsu
Graduate Institute of Science Education
National Taiwan Normal University
Taipei, Taiwan

ISBN 978-981-10-1184-9   ISBN 978-981-287-441-2   (eBook)
DOI 10.1007/978-981-287-441-2

Springer Singapore Heidelberg New York Dordrecht London

Printed on acid-free paper

Springer Science+Business Media Singapore Pte Ltd. is part of Springer Science+Business Media (www.springer.com)

Preface

Teaching with technology has been widespread around the world for many years, and it continues to increase and change with greater access to and variety of technologies. Taking the United States as an example, at least one computer was available in 97 % of its classrooms in 2009 (Gray, Thomas, & Lewis, 2010). The fever about pursuing electronic classrooms has spread to developing countries. Zhang (2007) reported that governments of most Eastern countries launched projects or reforms in classrooms since 2000 to infuse information and communication technologies (ICT). As early as 2002, Taiwan, Hong Kong, and three other Asian countries were listed among the countries with high ICT access (International Telecommunication Union, 2003; Ono, 2005). The Ministry of Education in Taiwan announced the *Blueprint of Information Education in Elementary and Junior High Schools* (Ministry of Education, 2008) to strengthen students' technology literacy, which included increasing the hours for information courses in the curriculum and enhancing teachers' technology application abilities. This call for technology literacy should not be confused with engineering/technology literacy advocated as part of the science, technology, engineering, and mathematics (STEM) initiative in the recent framework of science education in the USA (National Research Council [NRC], 2012; Yore, 2011); but ICT literacy is a contributing component to these science, mathematics, and engineering/technology literacies.

It appears that in Asia the technological hardware, software applications, and curricular authority are in place. Demands for teacher education to refine science teachers' technology usage are needed in Taiwan, Hong Kong, and elsewhere. Sung, Chang, and Hou (2005) pointed out that most of the teacher education courses in Taiwan regarding ICT literacy emphasized technological skills and lacked meaningful connections with pedagogy or subject area content. Practicing teachers reported positive attitudes toward technology implementation and knowledge about constructivist approaches as being effective strategies when teaching with technology, but their actual implementation was limited (Chen, 2008). As for the situation in Hong Kong, Fox and Henri (2005) found that teachers' teaching practices with technology changed little due to the administrative overloads and busy work even though the government introduced a policy that promoted a shift of curriculum and instruction to a student-centered focus. Gaps are common between teachers' actual implementation of technology and their knowledge levels or the expectations others place on them regarding teaching with technology. Some other problems that may discourage teachers' use of technology included the lack of teaching experience, negative perceptions of technology, insufficient on-site time, and technical or financial support (Mumtaz, 2000).

Clearly, helping teachers to teach with technology is critical to contemporary education since it will meet the needs of *Net Generation* students, capitalize on the potential of ICT to enrich and expand learning opportunities, and increase the effectiveness of the learning–teaching experience. Teachers' use of PowerPoint presentations to deliver their instruction in an organized way is a good first step to ICT-enriched instruction. However, there are still more profound possibilities where technology can be used to serve teaching and learning needs. Flexible uses of technology to realize content knowledge delivery and student learning are worth pursuing, such as through presenting natural phenomenon or allowing students to do simulated experiments in classrooms. It is only when teachers possess Technological Pedagogical Content Knowledge (TPACK; Mishra & Koehler, 2006) that would enable them to design and implement instruction with best considerations of students, curriculum, and technology. Assuming that the quality of teachers' technology utilization partially determines the quality of students' learning effects, then professional development of teachers' TPACK deserves a fuller investigation to identify what can be done better.

Purpose of the Book

The Science Education Center at National Taiwan Normal University has been exploring topics of science teaching and learning and designing technology-enabled science instruction for years. With these longitudinally academic research endeavors, the Center received grants from the Aim for the Top University Project that is funded by the Ministry of Science Education in Taiwan for making an overall improvement for science education in terms of science learning, teaching practice, policy, and research. This book tries to report on what the Center and its associates have done in promoting science teachers' instructional knowledge in teaching with technology and how they refine their science instruction with technological supports. Important TPACK issues, professional development, and teacher development are discussed, including theoretical and practical concerns, knowledge framework construction and evaluation rubrics, and actual observations on science teachers' TPACK development.

Organization of the Book

As Sir Isaac Newton wrote in a letter to Robert Hooke in 1676, "If I have seen further it is by standing on the shoulders of giants." Relying on the works of Shulman, Mishra, Koehler, Angeli, Valanides, and some other researchers, we propose a TPACK framework called TPACK-Practical (TPACK-P) within which teaching practices play key roles in contextualizing and evolving teachers' knowledge in teaching with technology. Some related empirical studies that we have done based on TPACK-P to investigate and develop science teachers' TPACK are reported in chapters. In order to make the book more diverse in its content, we invited some researchers who use an integrative TPACK framework to share with us their research findings in some chapters. Hopefully, with empirical studies based on two major but different TPACK frameworks, we can provide our readers a comprehensive understanding of the development of teachers' TPACK and stimulate further studies that inform teacher education programs.

Part I – TPACK in Teaching Practices

Part I reports how TPACK is epistemologically defined and actually shaped. Most of the current TPACK frameworks were heuristic based, that is, trying to find out what constitutes teachers' TPACK and to seek ways to further refine current teacher education. In this section, we are eager to unveil how TPACK is composed from practical teaching and learning contexts such as in science classrooms. We think sketches of the TPACK that teacher educators and science teachers have would inform the current development status of teachers' TPACK and provide insights about how we can work on and with it. In Chap. 1, we provide a brief overview introducing why TPACK is viewed as a strand of teachers' pedagogical content knowledge (PCK) and how the conceptualizations of the integrative frameworks and transformative frameworks inherited from PCK were instrumental in our view of TPACK. With the basic understanding of these two different schools of thought on the development of teachers' instructional knowledge, this chapter also raises the importance of *students* and *content* in teachers' TPACK development and classroom implementation. Chapter 2 begins with a synthesis of the frameworks and factors that we have known about TPACK based on previous research findings. The study that informed this chapter focused on knowing more about how TPACK is carried out within an actual teaching context. The authors invited a research panel and an expert teacher panel to participate in a Delphi survey to identify a practical framework of TPACK that teachers develop for and from actual teaching contexts. The authors in Chap. 3 documented inservice science teachers' use of technology when they designed their own curricula, enacted their teaching, and assessed their students' learning progress. Profiles of teachers who develop their TPACK-P with different proficiency levels are presented. Chapters 2 and 3 will familiarize readers about what TPACK-P in science teaching practices looks like and how we can strengthen science teachers' TPACK-P.

Part II – The Transformative Model of TPACK

The chapters in Part II take the transformative approach to view the development of teachers' TPACK. In Chap. 4, we try to construct rubrics that can be useful for evaluating preservice teachers' performances regarding lesson planning and microteaching in technology-infused classrooms. Though limited in number of participants, profiles of the seven preservice teachers' lesson plans and microteaching performances were carefully analyzed, suggesting directions for future teacher education. The longitudinal, complicated, and dynamic development varied for these TPACK exhibits, which makes knowledge measurement a difficult task. Findings from this chapter are expected to make contributions to the literature of TPACK measurements. In Chap. 5, the authors propose a teacher community of practice in which teachers with different proficiency levels in TPACK work together, not only for developing android applications (APPs) on multitouch tablets to facilitate students' physics learning but also for strengthening each others' TPACK-P for better accommodating student learning.

Part III – The Integrative Model of TPACK

Part III takes the integrative perspective to view the development of teachers' TPACK as an alternative perspective in applying TPACK in teacher education. Authors in Chap. 6 propose a model, MAGDAIRE (modeled analysis, guided development, articulated implementation, and reflected evaluation), to foster preservice teachers' ability to integrate ICT and teaching. They discuss how MAGDAIRE significantly increased preservice teachers' Flash knowledge and skills facilitated the development of their TPACK-Flash and led to better integration of the knowledge components. In Chap. 7, Hong Kong teachers' use of multimedia resources is analyzed using the TPACK framework to provide tangible understanding of how technology supports teaching and learning in elementary schools. Data collection included a revised questionnaire on TPACK and lesson observations of teachers' use of pedagogies with follow-up interviews, which provided some useful information to enhance our understanding of how multimedia resources are used in primary classrooms and initiated ideas about elementary teachers' teaching practices with technology.

Part IV – Epilogue

Chapter 8 provides external reflections and an international perspective to Chaps. 1, 2, 3, 4, 5, 6, and 7. Since this book discusses the development of teachers' TPACK in teaching practice, the Epilogue comments on how efficient these teachers or teacher educators were in the development of TPACK from epistemological and practical points of view. Insights for constructing a more teacher- and learner-friendly classroom with ICT implementation are offered.

Throughout this book, we would like to share with our readers what TPACK looks like when science teachers apply it in their teaching practice as well as what we have done in developing science teachers' TPACK from different perspectives. Perspectives and findings mainly from Taiwan and some from Hong Kong may not depict full pictures of the science teachers in the digital era in Asia or around the world. However, we expect these studies can be cases that give future researchers or educators insights about how educational negotiations between teaching and technology can be made. At the same time, we also look forward to more studies that investigate how science teachers can better teach with technological applications that include fundamental considerations of pedagogical concerns, science knowledge, and the dynamics and diversity of students and teachers. Technology can be constructive to student learning only if it is properly used and meaningfully engaged into instruction. Otherwise, technology can be a tool like blackboards that serve only as a knowledge displayer instead of a knowledge facilitator.

Acknowledgments

This book (Chaps. 1–5) was partially supported by the "Aim for the Top University Project" of National Taiwan Normal University (NTNU), sponsored by the Ministry of Education, Taiwan, R.O.C. The authors gratefully acknowledge the assistance of Prof. Larry Dean Yore and his wife, Sharyl Yore.

Taipei, Taiwan   Ying-Shao Hsu

References

Chen, C.-H. (2008). Why do teachers not practice what they believe regarding technology integration? *Journal of Educational Research, 102*(1), 65–75.

Fox, R., & Henri, J. (2005). Understanding teacher mindsets: IT and change in Hong Kong schools. *Educational Technology & Society, 8*(2), 161–169.

Gray, L., Thomas, N., & Lewis, L. (2010). *Teachers' use of educational technology in U.S. public schools: 2009* (NCES 2010–040). Washington, DC: National Center for Education Statistics, Institute of Education Sciences, US Department of Education.

International Telecommunication Union. (2003, November 19). *ITU digital access index: World's first global ICT ranking* [Press release].

Ministry of Education. (2008). *White Paper on information technology education for elementary and junior high schools 2008–2011*. Taipei, Taiwan: Author.

Mishra, P., & Koehler, M. J. (2006). Technological pedagogical content knowledge: A framework for teacher knowledge. *Teachers College Record, 108*(6), 1017–1054.

Mumtaz, S. (2000). Factors affecting teachers' uses of information and communications technology: A review of the literature. *Journal of Information Technology for Teacher Education, 9*(3), 319–342.

National Research Council. (2012). *A framework for K–12 science education: Practices, crosscutting concepts, and core ideas* (H. Quinn, H. A. Schweingruber, & T. Keller, Eds.). Washington, DC: National Academies Press.

Ono, H. (2005). Digital inequality in East Asia: Evidence from Japan, South Korea, and Singapore. *Asian Economic Papers, 4*, 116–139.

Sung, Y.-T., Chang, K.-E., & Hou, H.-T. (2005). Technology-instruction integration: Learning from America's experience and reflecting on Taiwan's development. *Bulletin of Educational Research, 51*(1), 31–62.

Yore, L. D. (2011). Foundations of scientific, mathematical, and technological literacies – Common themes and theoretical frameworks. In L. D. Yore, E. Van der Flier-Keller, D. W. Blades, T. W. Pelton, & D. B. Zandvliet (Eds.), *Pacific CRYSTAL centre for science, mathematics, and technology literacy: Lessons learned* (pp. 23–44). Rotterdam, The Netherlands: Sense.

Zhang, J. (2007). A cultural look at information and communication technologies in Eastern education. *Education Technology Research Development, 55*(3), 301–314.

Contents

Contributors

Chun-Yen Chang is a Professor of the Graduate Institute of Science Education at NTNU. He received his PhD from the University of Texas. Dr Chang served as the President of the National Association for Science Education in Taiwan (NASET) from 2011 to 2012 and now continues his work as the Associate Editor for the *Journal of Geosciences Education*. He is also on the Editorial Board of two SSCI-level journals: *Studies in Science Education* (science education) and *Learning, Media & Technology* (instructional technology) and served on the Editorial Board of *Journal of Research in Science Teaching* (science education) and as the Associate Editor for the *International Journal of Science and Mathematics Education* both from 2010 to 2013. Dr Chang has authored and coauthored more than 100 articles, including almost 80 papers indexed in the Science/Social Science Citation Index (SCI/SSCI) database. In 2003, 2009, and 2012, he received the Outstanding Researcher Award in 2003, 2009, and 2012 from National Science Council of Taiwan.

Sung-Pei Chien is a doctoral student in science education at National Taiwan Normal University, Taipei. His research interests include teacher education, teachers' beliefs, and teachers' community. His work has been published in *Computers in Human Behavior, Journal of Educational Computing Research, and International Journal of Science and Mathematics Education.* He also serves as a manuscript reviewer of *International Journal of Science and Mathematics Education.*

Yu-Ta Chien is a doctoral student in science education at the NTNU. He also serves as an engineer at the Science Education Center for developing e-learning environments. His research mainly focuses on using emerging technologies to facilitate science learning and teaching in the classroom. He has published several articles in international journals, such as *Computers & Education, British Journal of Educational Technology*, and *Teaching and Teacher Education.*

Fiona Ngai-Ying Ching is a PhD candidate at the Department of Science and Environmental Studies of The Hong Kong Institute of Education. Deeply interested in the brain, her research is focused on the preservice teachers' neuroscience literacy and perceptions of the application of neuroscience research in education with an aim to provide recommendations for making neuroscience part of initial teacher education. She is also interested in science education and ICT in education.

Apple Wai-Ping Fok is the Project Director of Pervasive Educational Technology Group, Centre for Innovative Applications of Internet and Multimedia Technologies (AIM*Tech* Centre), City University of Hong Kong; and the CEO of Sky Jump Enterprise Limited, a limited liability company incorporated in Hong Kong. She received her PhD in Computer Science specializing in personalized education from City University of Hong Kong. Her current R&D activities mainly focus on applying personalization technologies, intelligence agents, and semantic web technologies in education. Dr Fok created a full-scale e-education platform and deployed a blended learning curriculum framework with a unique e-pedagogical approach that enhances learning and teaching effectively.

Danah Henriksen is a Visiting Assistant Professor in the Educational Psychology and Educational Technology program at Michigan State University, East Lansing. She received her Ph.D. in educational psychology and educational technology from Michigan State University. Her research and scholarship has focused on creativity in teaching learning and the role that digital technologies can play in the process, specifically characteristics of the most effective and creative teachers, evaluation schemas for creative work, and transdisciplinary thinking. She coleads the College's Deep-Play Research Group, which focuses on research related to creativity, transdisciplinary thinking, and the twenty-first-century issues of teaching and learning. She serves as Assistant Chair of the Creativity SIG at the Society of Information Technology in Teacher Education.

Ying-Shao Hsu is an Adjunct Professor of Graduate Institute of Science Education and the Department of Earth Sciences at the National Taiwan Normal University (NTNU). Currently, she is also the director of Graduate Institute of Science Education at present and Research Chair Professor of NTNU. She received her PhD degree in 1997 from the Department of Curriculum and Instruction at the Iowa State University. Her expertise in research includes technology-assisted learning, inquiry learning of science, science curriculum design, and earth science education. She has published journal articles in leading journals of science education such as *Science Education, International Journal of Science Education, British Journal of Educational Technology, International Journal of Science and Mathematics Education*, and *Research in Science Education*. She currently serves the Editorial Board of *International Journal of Science and Mathematics Education* and as a reviewer for several SSCI journals. She is a recipient of Outstanding Research Award of the National Science Council in February 2012.

Fu-Kwun Hwang is a Professor of Department of Physics at the National Taiwan Normal University. He received his PhD degree in 1992 from the University of Maryland. By viewing that most high-school students in Taiwan did not like physics, he devoted himself in making science learning more fun and more friendly by applying physics concepts to nature phenomena or everyday life. With research interests on the use of computer tools for the conceptual understanding of physics, he has developed hundreds of physics-related java applets on the web to help students to play with physics models in the simulation and enjoy the fun of physics. The applets or the educational websites he constructed not only received honors from major educational institutes around the world but also be deemed useful among teachers and more than 88 mirror sites were created.

Tsung-Hau Jen is an assistant researcher at the Science Education Center at NTNU. He received his PhD in 2000 from NTNU with a specialization in large-scale assessments, and he develops innovative assessment theories, constructs online assessment platforms, and promotes innovative assessments in science classrooms at primary and secondary schools. Besides being an active author and reviewer for international journals, he is also a popular speaker who gives both talks and workshops on the topics of educational measurements for science education researchers and in-service teachers.

Tzu-Chiang Lin is a postdoctoral researcher of the Graduate Institute of Digital Learning and Education at the National Taiwan University of Science and Technology. He received his PhD in 2012 from the Department of Life Science at NTNU. His research interests include science education, science teacher education, and educational technology. He has published journal articles in the field of educational research, such as *International Journal of Science and Mathematics Education* and *Journal of Science Education and Technology*. He has also served as a paper reviewer for journals such as *Cyberpsychology, Behavior, and Social Networking* and *Computers & Education*.

Punya Mishra is Professor of Educational Technology at Michigan State University, East Lansing, where he directs and teaches in the Master of Arts in Educational Technology program. He received his Ph.D. in educational psychology from the University of Illinois, Urbana–Champaign. He chairs the Creativity SIG at the Society for Information Technology in Teacher Education and is former chair of the Innovation and Technology Committee of the American Association of Colleges of Teacher Education. He has worked extensively in the area of technology integration in teacher education which led to the development (in collaboration with Dr. M. J. Koehler) of the Technological Pedagogical Content Knowledge (TPACK) framework, which has been described as being "the most significant advancement in the area of technology integration in the past 25 years." He is nationally and internationally recognized for his work on the theoretical, cognitive, and social aspects related to the design and use of computer-based learning environments. He has received over $7 million in grants, has published over 50 articles and book chapters, and has edited two books.

Michael Wai-Fung Liu is currently the panel head of General Studies of Ying Wa Primary School and was a Project Officer of Department of Science and Environmental Studies of the Hong Kong Institute of Education in 2012. Mr Liu has been invited to share his experience in curriculum development and teaching of General Studies by the teacher education institute as well as the Education Bureau of HKSAR. He has a Bachelor of Education degree in 2007 from the Hong Kong Institute of Education and received his Master of Arts degree in Curriculum Development and Teaching of Liberal Studies in 2012 from The Chinese University of Hong Kong.

Winnie Wing-Mui So is the Head and Professor of Department of Science and Environmental Studies, as well as the Associate Dean of the Graduate School at the Hong Kong Institute of Education. She is also the Director of the Centre for Education in Environmental Sustainability. She received her PhD in science teaching and learning from the University of Hong Kong. Professor So has rich and diversified experiences associated with learning and teaching of science in primary and secondary classrooms as well as teacher education. She has published her research in the *Computers and Education, International Journal of Mathematics and Science Education, Educational Research Journal*, etc.

Hsin-Kai Wu is a Professor of Graduate Institute of Science Education at the National Taiwan Normal University. She has a MS degree in organic chemistry and received her PhD degree in science education from the University of Michigan. She has published journal articles on chemistry visualization, scientific inscriptions, learning technologies, and inquiry learning in leading journals of science education. Currently, she serves the Editorial Board of *Science Education* and *Journal of Research in Science Teaching*, and she is also an Associate Editor of the *Internal Journal of Science and Mathematics Education*. Dr Wu was the 2003 recipient of the Outstanding Dissertation Award from the US-based National Association for Research in Science Teaching (NARST). In 2004, her research work was recognized by Wu Da-Yu Memorial Award from the National Science Council in Taiwan. She also received the Early Career Award from NARST, which is giving annually to the early researcher who demonstrates the greatest potential to make outstanding and continuing contributions to research in science education. She is also a recipient of Outstanding Research Award of the National Science Council in February 2009 and 2014.

Yi-Fen Yeh is a project-appointed research scientist of Science Education Center at the National Taiwan Normal University. She received her PhD in 2010 from the Department of Curriculum and Instruction at Texas A&M University. Her research interests include technological pedagogical content knowledge (TPACK), technology-assisted learning, science reading, and scientific inquiry. She has published articles in international journals, such as *International Journal of Science Education*, the *Journal of Science Education and Technology*, and *School of Science and Mathematics*.

Part I
TPACK in Teaching Practices

Chapter 1
The Development of Teachers' Professional Learning and Knowledge

Ying-Shao Hsu

> *Technology is just a tool. In terms of getting the kids working together and motivating them, the teacher is most important.*
>
> Bill Gates

1.1 Background

Living in the world where technology keeps advancing, teaching with technology becomes a must-do for teachers to consider in their instruction of the *Net Generation*. It matters not only for how it helps students construct their current learning but also for how it reinforces citizen's technological literacy and drives technological advances forward. Facing students who are digital natives (getting used to new technology and the explosion of new information), teachers need to be smart about what and how technology-assisted instructional approaches are taken. Successful educational reforms in promoting teaching with technology cannot be achieved without teachers. Teachers are both the agents and the targets of change – leading, supporting, and infusing technology into their classrooms.

In countries where classrooms were supplied with educational technology like the USA, teachers' actual usage of projectors, interactive whiteboards, and digital cameras was low at 72 %, 57 %, and 49 %, respectively, among those who reported having access to the technology, and 40 % rated their or their students' use of computers during instruction as *often* while another 20 % rated their and students' use as *sometimes* (Gray, Thomas, & Lewis, 2010). Survey results from Project Tomorrow (2008), a series of surveys conducted annually since 2007, also revealed that both teachers' and students' technology uses for educational purposes were at a low level (e.g., using computers to type up worksheets or complete assignments). School principals expected newly hired teachers to be proficient in teaching with technology; parents expressed positive attitudes toward digital learning–teaching approaches (e.g., mobile learning tools). Other agents, such as students and aspiring

Y.-S. Hsu (✉)
Graduate Institute of Science Education, National Taiwan Normal University, Taipei, Taiwan
e-mail: yshsu@ntnu.edu.tw

© Springer Science+Business Media Singapore 2015
Y.-S. Hsu (ed.), *Development of Science Teachers' TPACK*,
DOI 10.1007/978-981-287-441-2_1

teachers, used social networking tools and discussion boards frequently to assist communications in informal learning (Project Tomorrow, 2012, 2013). All of these survey results indicate the urgent demand for teachers' professional knowledge and willingness to use technologies to assist their instruction.

Beyond pursuing the quantity of technology implementation in classrooms, more and more teachers and educators seek content knowledge instruction with appropriate technology implementation. Integrative curriculum, where the borders of subject content are broken, is now becoming a trend for educators to pursue as a near-future or long-term goal. For example, more and more schools in the USA participate in science, technology, engineering, and mathematics (STEM) programs. These programs encourage teachers in these four subject areas to work together at developing an integrative curriculum with the purpose to build up students' ability to solve complex interdisciplinary problems. The appropriateness of teachers' uses of technology to assist their instruction not only determines students' content knowledge comprehension but also develops students' technological literacy (National Research Council [NRC], 2012; Yore, 2011). All these demands would be bounded by teachers' knowledge about enriching curriculum and assisting students' learning with appropriate uses of technology (e.g., knowing what representations are good for teaching certain types of subject content).

In fact, teachers' knowledge is a complex construct blended with their longitudinal input of knowledge and experiences. Besides conflicts between new knowledge to old knowledge systems, teachers' knowledge grows even more complicated with personal experiences or diverse contextual confines. Though teachers' knowledge is personally and dynamically changing, many educational researchers still endeavor to determine what composes teachers' knowledge in instruction. Only when teachers' knowledge is unveiled can teacher education be more effectively designed and implemented. In the following pages, I present the components of teachers' professional knowledge through historical progression and from different perspectives within the past three decades. An introductory discussion is also made to seek future directions for teacher education studies.

1.2 Development of Pedagogical Content Knowledge

Effective teachers possess knowledge to structure and facilitate learning opportunities of specified knowledge in comprehensible ways for the intended learners. Shulman (1986) proposed a teacher knowledge framework that integrated these critical ideas called pedagogical content knowledge (PCK). This integrative framework was composed of teachers' professional knowledge about subject matters called content knowledge (CK) and knowledge of instruction called pedagogical knowledge (PK). Shulman (1987) described PCK as "the blending of content and pedagogy into an understanding of how particular topics, problems, or issues are organized, represented, and adapted to the diverse interests and abilities of learners, and presented for instruction" (p. 8). In other words, PCK refers to the *craft knowledge* of content, teaching, learning, and context that teachers rely on when they help their students to construct understanding of a specific idea or domain. The choice of craft knowledge emphasizes that this knowledge flows from experience and practice and not necessarily from theoretical sources.

Teaching is part of a complex decision-making process in which factors involved in the process of teaching and learning need to be considered carefully before, during, and after the actual event. These reflective practices require anticipatory knowledge for planning the learning–teaching experience, real-time evaluations on actions to monitor and adjust teaching, and post hoc reflections on actions to inform future teaching. For such a process requiring teachers' contemplation in different stages of instruction, Shulman (1987) suggested that PCK could be decomposed into (a) content knowledge, (b) general pedagogical knowledge, (c) curriculum knowledge, (d) knowledge of learners and their characteristics, (e) knowledge of educational contexts, and (f) knowledge of educational goals and intentions. Other researchers have added the importance of teachers' knowledge about context and school culture (Cochran, DeRuiter, & King, 1993; Grossman, 1990), teacher beliefs (Kagan, 1992; Veal, 2004), and experiences (van Driel, Verloop, & de Vos, 1998) to the listing of considerations in PCK. Among these components, CK is viewed as the prerequisite knowledge in teacher development (van Driel et al., 1998; van Driel, De Jong, & Verloop, 2002). Other components and experiences would be those that help teachers envision, enact, and realize instruction in accommodating ways to meet the needs of learners and address the realities of the context. Knowledge about students' misconceptions or alternative concepts would be part of the essence of teachers' TPACK after considering content with pedagogical concerns based on their longitudinal experiences.

Although PCK is viewed as the integration of the knowledge sets previously mentioned, some researchers have argued that PCK is a unified construct and is dynamically changing. Cochran et al. (1993) suggested that PCK should be renamed as pedagogical content knowing (PCKg) to capture the in-the-moment aspect since teachers' knowledge should not only be situated but also student centered and changeable. Gess-Newsome (1999) proposed a framework of PCK, a unitary holistic interpretation, in which teachers' knowledge of subject matter and pedagogy are contextually bounded and cannot be teased apart or deconstructed into individual components. Teachers' experiences become one major source, influencing the activation of necessary knowledge and reinforcement of their further development at the same time. Magnusson, Krajcik, and Borko (1999) suggested that science teachers need to develop knowledge of science curricula, understanding student science learning, instructional strategies, and assessment of scientific literacy and comprehension. All of these knowledge components are functionally and reciprocally nurtured and shaped by teachers' orientations toward and experiences with science teaching. Briefly, teachers' beliefs, values, attitudes, experiences, and teaching goals play important, fundamental roles that determine the development and transformation of teachers' PCK.

Both the integrative and transformative frameworks offer approaches to the development of teachers' PCK from different points of views. The integrative framework emphasizes and identifies the fundamental knowledge subsets contributing to the grand concept of PCK. This view of PCK, as a knowledge integration framework, has major influence on the design of most current teacher education systems since such knowledge can be easily carried out in different subsystems with courses delivered separately, that is, science content in academic department courses, pedagogy in general education courses, and science pedagogy in science education courses and science clinical experiences. However, based on teacher education graduates' criticism that their programs have been fragmented and lack practical relevance (Barone, Berliner, Blanchard, Casanova, & McGowan, 1996; Sandlin, Young, & Karge, 1992) and from a bottom-up point of view, it makes much more sense to view experienced teachers' PCK as the goal for preservice teachers to achieve in the early years of their teaching careers as they become inducted into the teaching profession. Actual teaching experiences and practices would transform teachers' CK and PK into unique PCK since factors of subject matter, individual student needs, curriculum goals, school climate, learning environment, and realities of time and classrooms are contextualized and demand careful considerations. Since preservice teachers' concerns in internship settings are frequently focused on issues of survival, CK, and performance evaluation but not PCK building, the transformative framework of PCK provides teacher educators a unitary holistic view for an organic process that needs time to develop and grow within the teaching contexts.

1.3   Development of Technological Pedagogical Content Knowledge

Discussions regarding PCK and teaching with technology date back to the early 1990s. Dwyer, Ringstaff, and Sandholtz (1991) identified five stages of teachers' evolution in using multimedia to assist their instruction (i.e., entry, adoption, adaptation, appropriation, and invention). Similar developmental progress (i.e., recognizing, accepting, adapting, exploring, and advancing) was also identified from teachers' learning about certain technology to the consolidation of their technological pedagogical content knowledge (TPACK; Niess et al., 2009). Factors that influenced teachers' achievements in teaching with technology include teachers' motivation and commitment, external supports, and access to technology (Hadley & Sheingold, 1993). It was not until a decade later when Pierson (2001) suggested the term *technological pedagogical content knowledge* to emphasize the importance of teachers' technological knowledge connected with the concept of teachers' PCK. Niess (2005) also proposed a similar idea by using the term *technology-enhanced PCK*. Both of these ideas emphasized the instructional use of technologies to improve the comprehensibility of target/abstract concepts, achievement of learning outcomes, and learning–teaching effectiveness.

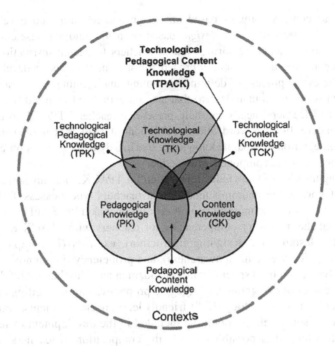

Fig. 1.1 TPACK framework (Koehler & Mishra, 2009, p. 63)

1.3.1   TPACK as an Integrative Framework

In order to decompose what contributes to teachers' knowledge in technology integration, Koehler and Mishra (2005) and Mishra and Koehler (2006) proposed a framework called *technological pedagogical content knowledge* (see Fig. 1.1). Similar to Shulman's idea that PCK is fundamentally an integrated body of knowledge composed of the intersection of CK and PK, TPACK refers to the overlapping area of CK, PK, and technological knowledge (TK). In fact, the composition of teachers' knowledge in technology integration was not as simple as merely adding TK into the PCK framework. The mutually integrated knowledge in the framework, including PCK, technological pedagogical knowledge (TPK), and technological content knowledge (TCK), also points out the knowledge that teachers need to engage in their instruction (Koehler & Mishra, 2009). Viewing TPACK as an extension of PCK, Graham et al. (2009) defined TPACK as the knowledge that teachers possess if they are able to know "a) how technological tools transform their pedagogical strategies and content representation for teaching particular topics, and b) how technological tools and representation impact a student's understanding of these topics" (p. 71).

Mishra and Koehler (2006) did not limit their consideration to technological innovations for instruction – again, not to be confused with engineering and technology as content domains based on design and research and development cycles. *Learning technology by design*, based on a longitudinal, research-based, teacher education method requiring student teachers to design instructional artifacts, has been used as the main approach for student teachers to develop their TPACK. The cyclic process of defining, designing, and refining the artifacts (e.g., instructional software, course design) that are contextualized within different subject topics and learners' needs can help preserve teachers' TPACK to develop, grow, and mature. In other words, the development of TPACK requires teachers to engage the integration of separate knowledge sets in a dynamic process for ensuring the interweaving of the component knowledge.

Following the idea of knowledge integration for TPACK, there are some important sets of knowledge or competencies that teachers who possess TPACK or teach effectively with technology should acquire. Kabakci Yurdakul et al. (2012) proposed that the necessary competencies for teachers with TPACK to develop included the design (i.e., designing instruction), exertion (i.e., implementing instruction), ethics (i.e., ethical awareness), and proficiency (i.e., innovativeness, problem solving, and field specializations). Guzman and Nussbaum (2009) stated that competencies of designing and engaging proper evaluations, setting information communication technology (ICT)-friendly learning environments, and retaining positive personal beliefs should be included in the development of teachers' TPACK. Therefore, it is possible to view the composition of teachers' TPACK with at least two tiers in the notion of knowledge integration. The integrative knowledge body of CK, PK, and TK offers the basis for teachers to carry out their instruction with technology while there are some intervening feedback loops during the enacted teaching that transform these knowledge components and their intersections.

1.3.2   TPACK as a Transformative Framework

Similar to the approaches taken to analyze teachers' PCK from an integrative framework, there are also transformative frameworks when considering the development of teachers' TPACK. Inherent in the transformative conceptualization of teacher education, which needs to be content specific, pedagogical, student centered, and situated (Cochran et al., 1993), the TPACK transformative framework is viewed as a unitary holistic body of knowledge that urges teachers to support content representations, learners, and pedagogy with careful consideration of the technological affordances (Angeli & Valanides, 2009). They proposed and defined the framework ICT–TPCK as:

> knowledge about tools and their pedagogical affordances, pedagogy, content, learners, and context are synthesized into an understanding of how particular topics that are difficult to be understood by learners, or difficult to be represented by teachers, can be transformed and taught more effectively with ICT, in ways that signify the added valued of technology. (Angeli & Valanides, 2009, pp. 158–159)

They claimed that functionality maps of instruction and available tools varied with situated contexts and individual students' learning progress. Under this rationale, TPACK is a knowledge construct that transforms with rounds of instructional decisions about what, who, and how to teach with careful consideration of the technological affordances and available technological resources.

The ways teachers' conceptualize TPACK determines the way their knowledge is developed. TPACK in an integrative framework assumes the additive components of knowledge where preservice teachers acquire subsets of knowledge that are applied to the design and development of instructional artifacts as practices for their TPACK integration. However, TPACK in a transformative framework assumes that teachers' TPACK transforms with experiences on designing and delivering content instruction with appropriate uses of technology and that it is the way teacher educators should follow to develop their students' TPACK. Angeli and Valanides, in a series of studies that explored formats for teachers developing TPACK, found that being proficient in one specific knowledge subset (e.g., CK, PK, TK, or TPK) would not ensure the likely development of TPACK (Angeli, 2005; Angeli & Valanides, 2005, 2009; Valanides & Angeli, 2006, 2008a, 2008b, 2008c). Full consideration of content, learners, and technological tools within the actual design would be one of the keys in teachers' TPACK transformation.

The emphasis of teacher education in a transformative framework is the mapping of knowledge of learners, pedagogy, representation, and tool affordances, as shown in Fig. 1.2 (Angeli & Valanides, 2009). This mapping, from constructivist learning theories, urges teachers to design instruction based on students' current knowledge (e.g., alternative concepts) and then engage, challenge, and arouse students' cognitive or sociocognitive conflicts to promote a conceptual change ecology. With this view, it is easier for in-service teachers to start from the alternative concepts that students might have by calling on their previous teaching experiences while preservice teachers would need to seek external information (e.g., experienced teachers' and teacher educators' guidance or information). Assuming that experienced teachers' PCK can be the guidance for preservice teachers, the idea of setting up communities of practice for preservice and in-service teachers where experienced teachers can demonstrate how they design and carry out the instruction could be another method to optimize the transformation of novice teachers' TPACK. According to the collaborative learning framework for teachers (Jang & Chen, 2010), either guided or self-initiated repetitive cycles of use, comprehension, observation, practice, and reflection in teaching with technology would be effective approaches to reinforce teachers' TPACK development in terms of transformative points of views.

Both integrative and transformative frameworks were deemed to be useful for teachers to help students acquire knowledge in a technology-enriched information age as one of the goals for teacher education. Seamless connections between technology and instruction are ideal goals for teachers rather than merely pursuing techno-centric classrooms (Ward & Kushner Benson, 2010). In that way, having a

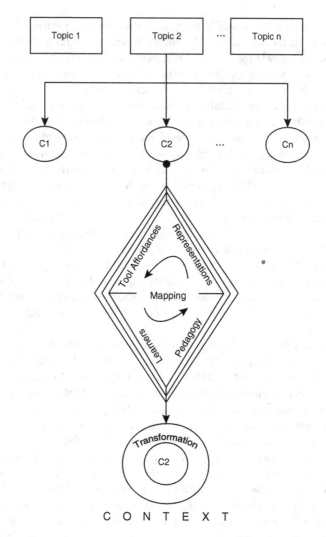

Fig. 1.2 ICT–TPCK in a situative instructional design model (Angeli & Valanides, 2009, p. 160)

technologically literate disposition or habit of mind is the basic requirement for teachers who will need to keep updating themselves so as to know how to use ever-changing hardware and software (Ertmer & Ottenbreit-Leftwich, 2010). The true value of TPACK would be ensuring more effective subject content instruction with consideration of pedagogical needs and use of appropriate technologies for accommodating students' learning needs. Since teacher knowledge is developed for and from teaching practices with technology in classrooms, the development of TPACK is an ongoing journey for preservice teachers, in-service teachers, and teacher educators.

1.4   TPACK as Twenty-First-Century Instructional Knowledge

TPACK is an integrated set of knowledge, but at the same time, transformation occurs within/between the component knowledge as well as to the overall construct. Some other descriptors like situated, dynamic, and multifaceted (Cox & Graham, 2009; Doering, Veletsianos, & Scharber, 2009; Koehler & Mishra, 2008) imply TPACK is a knowledge construct that varies and matures with contexts and is hard to be generally defined. No matter how complicated it may be in its development, it is undeniable that TPACK is born and elaborated for satisfying student learning needs. Who to be taught (i.e., students) has great influence on what to teach (e.g., subject content) and how to teach (e.g., pedagogy, technological tools) while how to teach will play a supportive role to the instructional target (i.e., learning goals). Furthermore, it becomes more complicated when TPACK is expected to be a knowledge framework that teachers rely on to develop students' twenty-first-century skills and competencies. Some researchers have proposed that the TPACK framework needs to be *transdisciplinary* (Kereluik, Mishra, & Koehler, 2010; Mishra, Koehler, & Henriksen, 2010). Besides CK learning, considering actual practices that teachers engage to enhance students' cognitive skills like problem solving and critical thinking is important for twenty-first-century education. By recapitulating the importance of accommodating student learning needs, TPACK should be student driven, content bound, and technology required but not technology prioritized.

Science teaching is a field that demands higher quality of teachers' TPACK. In contemporary views, science learning is involved with knowledge construction (Driver, Asoko, Leach, Scott, & Mortimer, 1994), conceptual change (Carey, 2000; Duit & Treagust, 2003), inquiry ability construction (Cuevas, Lee, Hart, & Deaktor, 2005; White & Frederiksen, 1998), and so on. Since misconceptions are common in science learning (Gil-Perez & Carrascosa, 1990; Gilbert & Watts, 1983), individual and social explorations of natural phenomena or data become necessary when students construct their science knowledge. Unobservable natural phenomena or abstract concepts have to be represented for students to visualize. Therefore, technology with different affordances can be helpful to student knowledge construction: multimedia for science phenomena presentation, predefined simulation software for students' modeling, and communication tools or platforms for collaborative learning. These science-specific disciplinary features make TPACK especially critical to science teachers since meaningful technological support and implementation can afford authentic science and engineering practices and scientific thinking in classrooms where the teacher-directed approach is pursued most of the time.

In the past few decades, science education researchers have devoted much time developing technologically assisted curricula or microcomputer-based experiments for helping students construct science CK and achieving scientific literacy. It was not until 2000 that researchers started paying attention to TPACK with regard to domain-specific content in science. Jimoyiannis (2010) in his technological pedagogical science knowledge (TPASK) pointed out content-specific knowledge components that science teachers are expected to develop, like "fostering scientific inquiry with ICT [and] student scaffolding" (p. 1263). Researchers spent more time on examining science teachers' proficiency in TPACK (Graham et al., 2009; Lin, Tsai, Chai, & Lee, 2013), but comparatively fewer studies have been conducted about the development of science teachers' knowledge and uses of certain technological devices (Jang & Tsai, 2012). Either specific competencies (e.g., science literacy, inquiry) or specific contents (e.g., plate movement, molecular collision) deserve quality technology-implemented instruction. Experienced science teachers' ideas can be good resources when designing learning tools or software that promotes students' science learning. Their meaningful and flexible uses of technology to assist science instruction can be practical exemplars for novice teachers to learn and observe. Networking among teacher communities comprised of individuals possessing varied scientific competencies can be another good approach to encourage sharing of science teachers' teaching materials in diverse formats and learning goals. Investigations on the nature and approaches to develop general TPACK (interdisciplinary) are important, but domain-specific TPACK within other disciplines should be emphasized as well in science education. After all, discussions become more consolidated once real teaching contexts (e.g., target content and students' learning progress) are considered.

References

Angeli, C. (2005). Transforming a teacher education method course through technology: Effects on preservice teachers' technology competency. *Computers & Education, 45*(4), 383–398.

Angeli, C., & Valanides, N. (2005). Preservice teachers as information and communication technology designers: An instructional design model based on an expanded view of pedagogical content knowledge. *Journal of Computer-Assisted Learning, 21*(4), 292–302.

Angeli, C., & Valanides, N. (2009). Epistemological and methodological issues for the conceptualization, development, and assessment of ICT-TPCK: Advances in technological pedagogical content knowledge (TPCK). *Computers & Education, 52*(1), 154–168.

Barone, T., Berliner, D. C., Blanchard, J., Casanova, U., & McGowan, T. (1996). A future for teacher education. In J. Sikula, T. Buttery, & E. Guyton (Eds.), *Handbook of research on teacher education* (2nd ed., pp. 1108–1149). New York: Macmillan.

Carey, S. (2000). Science education as conceptual change. *Journal of Applied Developmental Psychology, 21*(1), 13–19.

Cochran, K. F., DeRuiter, J. A., & King, R. A. (1993). Pedagogical content knowing: An integrative model for teacher preparation. *Journal of Teacher Education, 44*(4), 263–272.

Cox, S., & Graham, C. R. (2009). Using an elaborated model of the TPACK framework to analyze and depict teacher knowledge. *TechTrends, 53*(5), 60–71.

Cuevas, P., Lee, O., Hart, J., & Deaktor, R. (2005). Improving science inquiry with elementary students of diverse backgrounds. *Journal of Research in Science Teaching, 42*(3), 337–357.

Doering, A., Veletsianos, G., Scharber, C., & Miller, C. (2009). Using the technological, pedagogical, and content knowledge framework to design online learning environments and professional development. *Journal of Educational Computing Research, 41*(3), 319–346.

Driver, R., Asoko, H., Leach, J., Scott, P., & Mortimer, E. (1994). Constructing scientific knowledge in the classroom. *Educational Researcher, 23*(7), 5–12.

Duit, R., & Treagust, D. F. (2003). Conceptual change: A powerful framework for improving science teaching and learning. *International Journal of Science Education, 25*(6), 671–688.

Dwyer, D. C., Ringstaff, C., & Sandholtz, J. H. (1991). Changes in teachers' beliefs and practices in technology-rich classrooms. *Educational Leadership, 48*(8), 45–52.

Ertmer, P. A., & Ottenbreit-Leftwich, A. T. (2010). Teacher technology change: How knowledge, confidence, beliefs, and culture intersect. *Journal of Research on Technology in Education, 42*(3), 255–284.

Gess-Newsome, J. (1999). Pedagogical content knowledge: An introduction and orientation. In J. Gess-Newsome & N. G. Lederman (Eds.), *Examining pedagogical content knowledge: The construct and its implications for science education* (pp. 3–17). Dordrecht, The Netherlands: Kluwer.

Gilbert, J. K., & Watts, D. M. (1983). Concepts, misconceptions and alternative conceptions: Changing perspectives in science education. *Studies in Science Education, 10*(1), 61–98.

Gil-Perez, D., & Carrascosa, J. (1990). What to do about science "misconceptions". *Science Education, 74*(5), 531–540.

Graham, C. R., Burgoyne, N., Cantrell, P., Smith, L., St. Clair, L., & Harris, R. (2009). TPACK development in science teaching: Measuring the TPACK confidence of inservice science teachers. *TechTrends, 53*(5), 70–79.

Gray, L., Thomas, N., & Lewis, L. (2010). *Teachers' use of educational technology in U.S. public schools: 2009 (NCES 2010–040)*. Washington, DC: National Center for Education Statistics, Institute of Education Sciences, US Department of Education.

Grossman, P. (1990). *The making of a teacher: Teacher knowledge and teacher education*. New York: Teachers College Press.

Guzman, A., & Nussbaum, M. (2009). Teaching competencies for technology integration in the classroom. *Journal of Computer Assisted Learning, 25*(5), 453–469.

Hadley, M., & Sheingold, K. (1993). Commonalities and distinctive patterns in teachers' integration of computers. *American Journal of Education, 101*(3), 261–315.

Jang, S.-J., & Chen, K.-C. (2010). From PCK to TPACK: Developing a transformative model for pre-service science teachers. *Journal of Science Education and Technology, 19*(6), 553–564.

Jang, S.-J., & Tsai, M. F. (2012). Exploring the TPACK of Taiwanese elementary mathematics and science teachers with respect to use of interactive whiteboards. *Computers & Education, 59*(2), 327–338.

Jimoyiannis, A. (2010). Designing and implementing an integrated technological pedagogical science knowledge framework for science teachers' professional development. *Computers & Education, 55*(3), 1259–1269.

Kabakci Yurdakul, I., Odabasi, H. F., Kilicer, K., Coklar, A. N., Birinci, G., & Kurt, A. A. (2012). The development, validity and reliability of TPACK-deep: A technological pedagogical content knowledge scale. *Computers & Education, 58*(3), 964–977.

Kagan, D. M. (1992). Implication of research on teacher belief. *Educational Psychologist, 27*(1), 65–90.

Kereluik, K., Mishra, P., & Koehler, M. (2010). Reconsidering the T and C in TPACK: Repurposing technologies for interdisciplinary knowledge. In D. Gibson & B. Dodge (Eds.), *Proceedings of society for information technology & teacher education international conference 2010* (pp. 3892–3899). Chesapeake, VA: Association for the Advancement of Computing in Education (AACE).

Koehler, M. J., & Mishra, P. (2005). What happens when teachers design educational technology? The development of technological pedagogical content knowledge. *Journal of Educational Computing Research, 32*(2), 131–152.

Koehler, M. J., & Mishra, P. (2008). Introducing TPCK. In American Association of Colleges for Teacher Education Committee on Innovation and Technology (Ed.), *The handbook of technological pedagogical content knowledge (TPCK) for educators* (pp. 3–29). New York: Routledge.

Koehler, M. J., & Mishra, P. (2009). What is technological pedagogical content knowledge (TPACK)? *Contemporary Issues in Technology and Teacher Education, 9*(1), 60–70.

Lin, T. C., Tsai, C. C., Chai, C. S., & Lee, M. H. (2013). Identifying science teachers' perceptions of technological pedagogical and content knowledge (TPACK). *Journal of Science Education and Technology, 22*(3), 325–336.

Magnusson, S., Krajcik, J., & Borko, H. (1999). Nature, sources, and development of pedagogical content knowledge for science teaching. In J. Gess-Newsome & N. G. Lederman (Eds.), *Examining pedagogical content knowledge: The construct and its implications for science education* (pp. 95–132). Dordrecht, The Netherlands: Kluwer.

Mishra, P., & Koehler, M. J. (2006). Technological pedagogical content knowledge: A framework for teacher knowledge. *Teachers College Record, 108*(6), 1017–1054.

Mishra, P., Koehler, M. J., & Henriksen, D. (2010). The 7 trans-disciplinary habits of mind: Extending the TPACK framework towards 21st century learning. *Educational Technology, 51*(2), 22–28.

National Research Council. (2012). In H. Quinn, H. A. Schweingruber, & T. Keller (Eds.), *A framework for K–12 science education: Practices, crosscutting concepts, and core ideas.* Washington, DC: National Academies Press.

Niess, M. L. (2005). Preparing teachers to teach science and mathematics with technology: Developing a technology pedagogical content knowledge. *Teaching and Teacher Education, 21*(5), 509–523.

Niess, M. L., Ronau, R. N., Shafer, K. G., Driskell, S. O., Harper, S. R., Johnston, C., et al. (2009). Mathematics teacher TPACK standards and development model. *Contemporary Issues in Technology and Teacher Education, 9*(1), 4–24.

Pierson, M. (2001). Technology integration practice as a function of pedagogical expertise. *Journal of Research on Computing in Education, 33*(4), 413–430.

Project Tomorrow. (2008). *21st century learners' deserve a 21st century education: Selected national findings of the speak up 2007 survey.* Retrieved from http://www.tomorrow.org/docs/national%20findings%20speak%20up%202007.pdf

Project Tomorrow. (2012). *Learning in the 21st century: Digital experiences and expectations of tomorrow's teachers.* Retrieved from http://www.tomorrow.org/speakup/tomorrowsteachers_report2013.html

Project Tomorrow. (2013). *2013 trends in online learning virtual, blended and flipped classrooms.* Retrieved from http://www.tomorrow.org/speakup/2013_OnlineLearningReport.html

Sandlin, R. A., Young, B. L., & Karge, B. D. (1992). Regularly and alternatively credentialed beginning teachers: Comparison and contrast of their development. *Action in Teacher Education, 14*(4), 16–23.

Shulman, L. S. (1986). Those who understand: Knowledge growth in teaching. *Educational Researcher, 15*(2), 4–14.

Shulman, L. S. (1987). Knowledge and teaching: Foundations of the new reform. *Harvard Educational Review, 57*(1), 1–22.

Valanides, N., & Angeli, C. (2006). Preparing preservice elementary teachers to teach science through computer models. *Contemporary Issues in Technology and Teacher Education – Science, 6*(1), 87–98.

Valanides, N., & Angeli, C. (2008a). Distributed cognition in a sixth-grade classroom: An attempt to overcome alternative conceptions about light and color. *Journal of Research on Technology in Education, 40*(3), 309–336.

Valanides, N., & Angeli, C. (2008b). Learning and teaching about scientific models with a computer modeling tool. *Computers in Human Behavior, 24*, 220–233.

Valanides, N., & Angeli, C. (2008c). Professional development for computer-enhanced learning: A case study with science teachers. *Research in Science and Technological Education, 26*(1), 3–12.

van Driel, J. H., De Jong, O., & Verloop, N. (2002). The development of preservice chemistry teachers' pedagogical content knowledge. *Science Education, 86*(4), 572–590.

van Driel, J. H., Verloop, N., & de Vos, W. (1998). Developing science teachers' pedagogical content knowledge. *Journal of Research in Science Teaching, 35*(6), 673–695.

Veal, W. R. (2004). Beliefs and knowledge in chemistry teacher development. *International Journal of Science Education, 26*(3), 329–351.

Ward, C. L., & Kushner Benson, S. N. (2010). Developing new schemas for online teaching and learning: TPACK. *MERLOT Journal of Online Learning and Teaching, 6*(2), 482–490. Retrieved from http://jolt.merlot.org/vol6no2/ward_0610.htm

White, B. Y., & Frederiksen, J. R. (1998). Inquiry, modeling, and metacognition: Making science accessible to all students. *Cognition and Instruction, 16*(1), 3–118.

Yore, L. D. (2011). Foundations of scientific, mathematical, and technological literacies – Common themes and theoretical frameworks. In L. D. Yore, E. Van der Flier-Keller, D. W. Blades, T. W. Pelton, & D. B. Zandvliet (Eds.), *Pacific CRYSTAL centre for science, mathematics, and technology literacy: Lessons learned* (pp. 23–44). Rotterdam, The Netherlands: Sense.

Chapter 2
The TPACK-P Framework for Science Teachers in a Practical Teaching Context

Ying-Shao Hsu, Yi-Fen Yeh, and Hsin-Kai Wu

TPACK refers to the knowledge construct that teachers rely on to facilitate their instruction with technology. In order to decompose what constitutes this knowledge construct, researchers have proposed and validated frameworks from different perspectives or for different purposes. However, no one has tried to develop a working model of TPACK within an actual teaching context such as science. Therefore, we recruited experts and experienced science teachers to participate in panels and used the Delphi survey technique to collect their ideas and develop consensus for the framework of TPACK-Practical (TPACK-P) that reflects how teachers applied TPACK while teaching science in their classrooms. A total of eight knowledge dimensions were identified as critical contributions to science teachers' TPACK-P; 17 indicators were generated to further define the specifics of these knowledge dimensions. This framework of TPACK-P will give novice science teachers ideas about expert science teachers' technology-infused instructional practices and inform science teacher educators about critical technological aspects that should be facilitated in science teacher education programs.

2.1 Introduction

Technologies (e.g., overhead projectors, televisions, computers, and numerous specialized devices) have been used and misused in classrooms over the last 50 years. However, the explosion of modern microelectronic information communication technologies (ICTs) marked a new potential for using these technologies in science

Y.-S. Hsu (✉) • H.-K. Wu
Graduate Institute of Science Education, National Taiwan Normal University, Taipei, Taiwan
e-mail: yshsu@ntnu.edu.tw

Y.-F. Yeh
Science Education Center, National Taiwan Normal University, Taipei, Taiwan

© Springer Science+Business Media Singapore 2015
Y.-S. Hsu (ed.), *Development of Science Teachers' TPACK*,
DOI 10.1007/978-981-287-441-2_2

classrooms. This potential and the current use of technologies in teaching has stimulated several researchers to start paying attention to the pedagogical and content
knowledge about teaching, learning, and assessment facilitated with technology.
Technological pedagogical and content knowledge (TPACK, originally called
TPCK) refers to a strand of pedagogical content knowledge (PCK) that teachers
developed during content-specific teaching with technology (Koehler & Mishra,
2005; Mishra & Koehler, 2006). TPACK is important because it refers to an underlying framework that teachers use when planning, enacting, and adjusting their
instruction to be more comprehensible to their students through the use of ICTs.

The primary duty of teachers is to help students acquire subject knowledge,
which is also the main goal when explaining why teachers need to develop
TPACK. Science is the subject area that extensively requires teachers to display
scientific phenomena that are usually abstract or hard to visualize as well as engage
friendly and safe environments for students to inquire and communicate ideas.
Therefore, it is important for science teachers to develop TPACK, especially about
the use of ICTs since ICT diversifies the types of representations they can use to
display and offers communication channels that might accommodate various
learning needs. Therefore, this chapter aims to identify the framework and the key
knowledge of science teachers' TPACK that science teachers in the Information Age
should develop.

2.1.1   What Is Lacking in TPACK?

TPACK refers to an integrated set of knowledge consisting of content knowledge
(CK), pedagogical knowledge (PK), and technological knowledge (TK). Mishra
and Koehler (2006) used a Venn diagram showing TPACK as the intersecting area
of three independent types of knowledge (i.e., TK, PK, CK) or the three intersections of two mutually integrated types of knowledge (i.e., TPK, TCK, TPCK).
Conceptualizing TPACK as the integration of different knowledge combinations
offers a flexible way to explain how teachers' knowledge assists their instruction for
different situations. For example, science teachers would need TPK to set up discussion boards for accommodating students' informal learning about socioscientific
issues; at other times, they need PCK to design and deliver instructions based on the
nature of science concepts and the students' prior knowledge about the concept.

It is easier for us to understand the composition of teachers' TPACK as the overlapping area where three facets of knowledge converge, but such a claim lacks supportive research findings. The component knowledge of PCK (i.e., PK, CK, and
PCK; Shulman, 1986) has already been found difficult to be validated as three distinctive categories of knowledge (McEwan & Bull, 1991; Segall, 2004), not even
mentioning the situations for the composition of TPACK. In a study requiring teachers to rate their proficiency in each knowledge subset and analyze what contributes
to their TPACK through think-aloud, Archambault and Crippen (2009) found high
correlations between CK and PK as well as teachers' inability to identify differences between them. Such results were suspected to be the ambiguous distinctions

between CK and PK. In a later study in which factor analyses were used to determine the component knowledge within the construct of TPACK, Archambault and Barnett (2010) only confirmed the notion that TPACK was composed of three facets of knowledge and that TK was distinctively different from CK and PK. Therefore, the fuzzy definitions and the lack of a firm theoretical foundation and stable construct validity would be the theoretical concerns to rationalize TPACK merely as an integration of knowledge combinations (Cox, 2008; Gess-Newsome, 2002; Graham, 2011; Magnusson, Krajcik, & Borko, 1999).

Rather than taking the traditional integrative view of TPACK in which CK is viewed as the intersections of three fundamental knowledge subsets, Angeli and Valanides (2009) argued that TPACK is a content-specific, holistic body of knowledge. They proposed the framework *ICT-TPCK* as a way to encompass teachers' knowledge composed of knowledge about ICT, pedagogy, content, learners, and context. Teachers' instructional designs should emphasize learner-centered approaches and the mapping of content representations with tool affordances within varying situated contexts and levels of students' learning progress. The development of teachers' TPACK is transformative-based since negotiations among knowledge subsets are unavoidable in each instructional design or act. Explicit guidance helping teachers to teach with technology, rather than developing knowledge subsets individually, was found to greatly benefit the quality of teachers' instruction. Although the transformative framework can be used to better address the dynamic and transactional interactions and connections among content, pedagogy, and technology, it is not an easy task for preservice teachers to conceptualize what TPACK is really about, especially when the current TPACK frameworks only discuss teacher knowledge composition epistemologically.

2.1.2 TPACK in a Practical Sense

The development of PCK and TPACK should not be viewed simply as either the integration or the transformation of different sets of knowledge (see Chap. 1 for a more complete comparison). Even through activities such as engaging preservice teachers to design instructional artifacts that researchers of integrative and transformative frameworks mainly adopt, the TPACK that preservice teachers develop is not the same body of knowledge that experienced teachers possess. First, the model of PCK emphasizes the dynamic nature of knowing in teachers' knowledge (Cochran, DeRuiter, & King, 1993). Knowledge is not static; it is growing whenever similar ideas are added or changing whenever new stimulus or discrepant knowledge is accommodated. Second, teaching experience should be viewed as the major resource of and influence on PCK, with CK and PK as prerequisites (van Driel, Verloop, & de Vos, 1998). Arranging instructional content to fulfill teaching goals for specific learners with appropriate strategies renders practical knowledge and PCK to interact, influence, and develop each other. These dynamic features make it difficult to deconstruct and separate the composite knowledge into components. Therefore, the importance of practical knowledge should not be deemphasized in the development of TPACK.

Practical knowledge is critical to teachers' education and development since it guides their decisions, actions, and practices and influences their knowledge with reflections on practice. van Driel, Beijaard, and Verloop (2001) described practical features as action-oriented, person-bounded and context-bounded, tacitly developed, multiknowledge integrated, and greatly influenced by teachers' beliefs. The complexity of what happens during instruction, especially teacher–student interactions (Feiman-Nemser & Remillard, 1996), is the nature of practical knowledge.

2.2 TPACK-Practical (TPACK-P)

Contextualized performances illustrate TPACK as a dynamic knowledge construct within realistic teaching situations and help close the theory–practice gap in teaching with technology. TPACK in a practical orientation can be used to delineate the TPACK that in-service teachers possess from years of teaching practice and that preservice teachers need to develop and document as consistently pursued goals. Therefore, we proposed a framework of unique knowledge that evolved from science teachers' TPACK and enriched with their practical experiences as *TPACK-Practical* (TPACK-P; Yeh, Hsu, Wu, Hwang, & Lin, 2014). In order to identify what may constitute appropriate TPACK-P for these in-service teachers, we collected ideas from educational researchers' perspectives and expert teachers' ideas. Communications and negotiations with and amongst these informants were engaged to reach a consensus knowledge framework using the Delphi survey technique.

2.2.1 Delphi Survey

The Delphi method is a systematic deliberation technique in which researchers collect expert panels' opinions; it enables experts to communicate anonymously with one another and then the researchers explore the underlying information collected about focus ideas or issues (Turoff, 1970). Delbecq, van de Ven, and Gustafson (1975) stated that the Delphi survey technique is "a method of systematic solicitation and collection of judgments on a particular topic through a set of carefully designed sequential questionnaires, interspersed with summarized information and feedback of opinions derived from earlier responses" (p. 10). The classic Delphi survey technique usually takes four to seven rounds to collect and communicate opinions (Young & Hogben, 1978), but flexible procedures with at least two rounds of opinion collection and deliberations were used in recent studies (Hasson, Keeney, & McKeena, 2000). Cochran (1983) suggested that panel sizes of at least ten experts were needed to maintain heterogeneity and homogeneity of the participants' backgrounds. For example, Osborne, Collins, Ratcliffe, Millar, and Duschl (2003) encouraged participating expert teachers to brainstorm what should be included in contemporary science curricula. The initial round of brainstorming had no agreement but, during successive rounds, the same group of experts modified and reached consensus through anonymous communications. Tigelaar, Dolmans, Wolfhagen, and van der Vleuten (2004) proposed a framework about teaching competencies and validated it with two rounds of formal survey with an expert panel. The number of survey rounds depends on how fast agreement amongst the experts is reached.

In this study, a research panel and an expert panel were recruited for framework generation and validation, respectively, in order to secure the framework to be academically and practically accepted. The Delphi method in this study involved three unique steps: the research panel proposed a target version of the focus idea (TPACK-P), then two rounds of two different expert panels' deliberations and comments about necessary revisions were engaged to seek consensus views and make necessary adjustments to the framework.

2.2.2   Framework Generation

First, the research panel proposed and drafted a preliminary framework of TPACK-P. The research panel consisted of six experts in educational technology, including three professors with over 10 years' experience in science teacher professional development in e-learning classrooms, two postdoctoral researchers, and one doctoral student. They drafted the temporary focus framework of TPACK-P and developed associated survey questions based mainly on the orientation to teaching science (Magnusson et al., 1999), which proposed that science teachers' PCK was composed of knowledge of (a) science curricula, (b) students' understanding of science, (c) instructional strategies, and (d) assessment of scientific literacy. In the model of pedagogical reasoning and action, Shulman (1987) pointed out that most teaching processes were initiated from the teachers' content comprehension and then moved to consider curriculum transformation, teaching acts, student evaluation, teacher reflection, and ended with the teachers enjoying a new and greater level of comprehension. Figure 2.1 shows the temporary focus framework of TPACK-P. The framework identified three major domains of knowledge with an ICT focus: learners, curriculum design, and classroom instruction. (Knowledge of learners at the top and at the bottom refers to the same knowledge set that teachers use to evaluate students, indicated by asterisks.) Knowledge of curriculum design has four subdimensions and knowledge of classroom instruction has two subdimensions.

These essential elements of teachers' knowledge about teaching with ICT can also be viewed as sequential stages that teachers engage, recursively and repeatedly, when enacting and completing instructional units. Dimensions A (using ICT to understand students) and H (assessment) refer to the knowledge that teachers use to ensure their curriculum design and classroom instruction are effective by understanding the initial learning status of their students and their learning progress. Both are assessment knowledge in TPACK-P, but we separate them because these

Fig. 2.1 The original
framework of TPACK-P to
stimulate deliberations

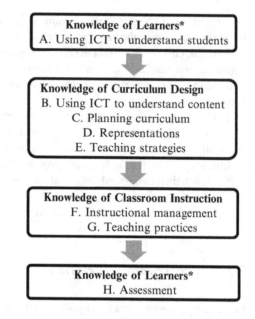

knowledge domains are also seen as sequential stages in instruction. Understanding their learners allows teachers to be better able to inform and demonstrate their curriculum-design knowledge through a linear but recursively sequential use of information from Dimension B (using ICT to understand content), Dimension C (planning curriculum), Dimension D (representations), and then Dimension E (teaching strategies). In the classroom teaching context, practical teaching was presupposed to include teachers' knowledge of Dimension F (instructional management) and Dimension G (teaching practices). Specific indicators for each knowledge dimension (a total of 20 indicators) were generated during the research panel discussions over three meetings.

2.2.3 Framework Validation

The temporary focus framework and indicators were submitted to the expert panel of informed practitioners to collect professional insights for the model evaluation, refinement, and expansion. The experts were defined in this study as faculty members teaching science in colleges or high schools in Taiwan, none of whom were involved in the research panel. They had specialties in science education or e-learning, 5 years of teaching experience with technology, teaching awards or instructional software design awards, or had actively participated in workshops and teacher groups focused on technology-infused instruction. Since the Delphi surveys were conducted in two rounds of anonymous deliberations and communications, only the experts who participated in both rounds and offered feedback were considered to be valid participants for the final analyses. The 54 experts participating in both rounds included 15 college faculty members and 39 science teachers (i.e., 8 earth science, 9 chemistry, 11 biology, and 11 physics teachers).

The survey formats for the two rounds of deliberations used a 5-point Likert scale for the expert faculty members and science teachers to rate their importance of the dimensions and indicators; blank spaces located beneath each statement were for them to provide feedback regarding item modifications as well as to posit their ideas about other possible knowledge dimensions or indicators. Experts in Round 1 received a survey packet that inquired about their ratings and comments to every item in the temporary TPACK-P framework. In the Round 2 survey, they received a survey packet about the modified knowledge dimensions and indicators based on Round 1 deliberations and responses, accompanied by an overall summary of the importance ratings and responses for all experts from Round 1 for their reference.

The importance of the eight knowledge dimensions in Rounds 1 and 2 was ranked in terms of either as 4 (important) or 5 (very important) by the majority of participants, which means that these original knowledge dimensions were viewed as critical and valid components to TPACK-P. A high consensus rating (96 %) of either 4 or 5 was found on the experts' ratings for each of the eight knowledge dimensions and indicators. The stability of rating importance on each item was in the range of 50–74 % through Rounds 1 and 2. Modifications on the knowledge dimensions were mainly for clarification and coherent connections among dimensions within the scenarios of technology-infused instruction. Modifications for the indicators included removal, addition, or combination of indicators depending on the qualitative comments given by experts (details of expert ratings, comments, and modifications are presented in Yeh et al., 2014). Table 2.1 shows the finalized version of the framework and the descriptive statistics of the importance rankings from the expert panel members.

2.3   The Framework of TPACK-P

The verified TPACK-P framework is composed of three main domains of knowledge about learners, curriculum design, and classroom instructions. Teachers develop preliminary TPACK (or PCK) before they are on board, but their prototypical knowledge would be further elaborated and become TPACK-P with more practical experience (e.g., making pedagogical decisions on usage and adjustment of teaching strategy, utilization of technology-infused tasks, facing difficulties on classroom management, etc.). In other words, the essence of TPACK-P is the craft knowledge that teachers develop for and from actual instruction with technology. The more classroom teaching and thoughtful, reflective practice that teachers engage, the more elaborated TPACK-P they can develop and the better technology-infused instruction they can offer. The eight dimensions that were identified by science education experts informed us of the key aspects that successful technology-infused instruction would need; the associated indicators are what teachers need to achieve to ensure that these aspects are pursued qualitatively.

Table 2.1 Descriptive statistics of experts' importance ratings regarding knowledge dimensions and indicators for the TPACK-P from Round 2 of the Delphi survey

Dimensions and Indicators of Information Communication Technologies (ICT)	Mean	Mode	SD
*Knowledge of Learners**			
A. Using ICT to understand students	**4.48**	**4**	**0.50**
A-1. Know how to use ICT to know more about students	4.17	4	0.54
A-2. Know how to use ICT to identify students' learning difficulties	4.65	5	0.52
A-3. Be able to use different technology-infused instruction to assist students with different learning characteristics	4.57	5	0.50
Knowledge of Planning and Designing			
B. Using ICT to understand subject content	**4.65**	**5**	**0.51**
B-1. Be able to use ICT to better understand the subject content	4.52	5	0.53
B-2. Be able to identify the subject topics that can be better presented with ICT	4.78	5	0.40
C. Planning ICT-infused curriculum	**4.41**	**4**	**0.59**
C-1. Be able to evaluate factors that influence the planning of ICT-infused curriculum	4.52	5	0.54
C-2. Be able to design technology-infused lessons or curriculum	4.71	5	0.49
D. Using ICT representations to present instructional representations	**4.60**	**5**	**0.52**
D-1. Be able to use appropriate ICT representations to present instructional content	4.76	5	0.43
E. Employing ICT-integrated teaching strategies	**4.48**	**5**	**0.53**
E-1. Be able to indicate the strategies that are appropriate to be used with ICT-integrated instruction	4.46	4	0.54
E-2. Be able to apply appropriate teaching strategies in technology-integrated instruction	4.69	5	0.46
Knowledge of Classroom Instruction			
F. Applying ICT to instructional management	**4.14**	**4**	**0.55**
F-1. Be able to use ICT to facilitate instructional management	4.18	4	0.57
G. Infusing ICT into teaching contexts	**4.54**	**5**	**0.57**
G-1. Be able to indicate the differences between the contexts of ICT-infused teaching to the contexts of traditional teaching	4.55	5	0.57
G-2. Be able to indicate the influences of different ICT to instruction	4.43	5	0.60
G-3. Be able to indicate substitute plans for technology-infused instruction	4.57	5	0.65
*Knowledge of Learners**			
H. Using ICT to assess students	**4.43**	**5**	**0.60**
H-1. Know the types of technology-infused assessment approaches	4.33	4	0.58
H-2. Be able to identify the differences between technology-integrated assessments to traditional assessments	4.34	4	0.47
H-3. Be able to use ICT to assess students' learning progress	4.55	5	0.56

2.3.1   Knowledge of Learners

It is good to initiate and close instruction with knowing about learners, such as knowing students' starting points and assessing their learning progress. In science learning, students may have a variety of misconceptions about science and specific topics and encounter difficulties during science learning (De Jong, van Driel, & Verloop, 2005; Jones & Moreland, 2005; Thompson, Christensen, & Wittmann, 2011). Knowing about learners prior to, during, and after instruction is deemed fundamental to quality instruction since it allows teachers to not only make appropriate and timely provisions in accordance with students' progress but also examine and reflect on their instruction (Davis & Krajcik, 2005; McNair, 2004; Otero, 2006). Interchangeable uses of formative assessments and summative assessments offer teachers flexible, alternative formats to track students' learning and to address the accountability responsibility placed on science teachers by school requirements.

The TPACK-P framework divided knowledge of learners into two parts: knowing how to use ICT to understand students (Dimension A) and knowing how to use ICT to assess students (Dimension H). The former focuses on the preassessment while the latter focuses on formative and summative assessments. Among the three indicators for using ICT to understand students (Dimension A), high importance scores were found in knowing how to use ICT to identify students' learning difficulties (A2, $M=4.65$) and being able to use different ICT-infused instruction to assist students with different learning characteristics (A3, $M=4.57$). Similar high importance was assigned to being able to use ICT to assess students' learning progress (H3, $M=4.55$). These three indicators focus on teachers' knowledge in being aware of science learners' cognitive development and using appropriate ICT to assist students based on what they know. In contrast, the comparatively lower importance ratings found for the indicator A1 ($M=4.17$), the knowledge about using ICT to know general aspects of students (e.g., affective states or learning preferences), indicated less priority. The high ratings on the identification of students' learning difficulties and assisting students' individual learning progress with ICT imply that these expert science teachers noticed the importance of alternative concepts and conceptual changes in science learning and viewed ICTs as diagnosis or assisted tools.

Science is a subject where misconceptions are common and scientific thinking can be complex for students to acquire in a short time. The variety of ICTs can be tools for teachers to assist the assessments they design or use to determine students' learning but not solutions. For example, online tests and simulation-based manipulations are good tools to measure students' fact knowledge and scientific thinking, respectively, but are not perfect if they are used alternatively. Therefore, science teachers need to ensure that they know how to act on their assessment knowledge and to evaluate students at the right time with the right tool(s) or in the right format.

2.3.2 Knowledge of Planning and Designing

Curriculum planning and designing is a complex thinking process – much like engineering and technological innovation – that requires teachers to anticipate all kinds of situations that may happen in teaching. Harris and Hofer (2009) stated that, "teachers' planning – which expresses teachers' knowledge-in-action in pragmatic ways – is situated, contextually sensitive, routinized, and activity-based" (p. 1). Pedagogical decisions are the product of teachers' considerations of institutional constraints, antecedent conditions, teacher characteristics, and cognitive processes that have consequences for teachers and students (Shavelson & Stern, 1981). Within the realm of science teaching, teachers' scientific thinking and knowledge of the nature of science are fundamental in determining how to design and to enact their instruction (Duschl & Wright, 1989). The variety of ICT applications (e.g., information searching, communication, and data analysis software) can be useful for teachers when they plan and design science learning activities.

In this study, teachers' knowledge about curriculum planning and designing was viewed as a transitioning bridge connecting what teachers know about their students and what teachers and students enact in the classroom. Four domains of knowledge (see Table 2.1) were identified as being needed when teachers design science curriculum, including using ICT to understand subject content (Dimension B), planning ICT-infused curriculum (Dimension C), using ICT representations to present instructional representations (Dimension D), and employing ICT-integrated teaching strategies (Dimension E). All the indicators for these dimensions were assigned comparatively high importance ratings except for being able to indicate the strategies that are appropriate to be used with ICT-integrated instruction (E1, $M=4.46$). Especially high ratings were found on identifying subject topics that can be better presented with ICT (B2, $M=4.78$), designing ICT-infused lessons or curriculum (C1, $M=4.71$), presenting instructional content with appropriate ICT representations (D1, $M=4.76$), and applying appropriate teaching strategies (E2, $M=4.69$). These indicators can be categorized into two groups: using the affordances of ICT to deliver content more effectively and ensuring the quality of ICT-infused instruction.

One major feature of ICT that benefits science content instruction is its potential to produce diverse representations for effective science teaching and learning, which frequently involve multimedia (e.g., text, sound, graphics, animation, videos) and data collecting and analyzing tools (e.g., simulations and modeling tools, microcomputer-based laboratories). Representations can be in multiple forms that facilitate science learners' visualization of microlevel or macrolevel phenomena and their understanding of scientific explanations or abstract concepts (Ainsworth, 2006; Mayer, 1999; Treagust, Chittleborough, & Mamiala, 2003; Wu, Krajcik, & Soloway, 2001). Dynamic representations should be a good way to present sequential phenomena whereas static representations should be a good way to present information that requires students to pay extra attention and scrutinize these data closely (Kozma, 2003). ICT-infused classrooms easily achieve these desired learning outcomes. Therefore, knowing what representations to use for certain subject content or contexts would demand teachers' pedagogical consideration; research studies have provided some insights. For example, learners acquire an abstract concept more effectively if graphical representations are presented prior to print-based text (Verdi, Johnson, Stock, Kulhavy, & Whitman-Ahern, 1997); presenting static representations first followed with dynamic representations facilitates students' conceptual understanding (van der Meji & de Jong, 2006; Wu, Lin, & Hsu, 2013); and better learning effects results would be attained when information is presented in combinational formats of representations that carry information in different channels (Mayer, 2009). Briefly, knowledge about representations involves knowing what types of subject content goes best with certain types of representations and knowing how to effectively deliver the ICT-infused content.

Technological tools can also be implemented as one part of the course curriculum or supplementary learning. Since developing students' inquiry ability is one major teaching goal for science teachers (National Research Council, 2012), ICT has been used to assist students' science learning by offering scaffolding and prompts (Quintana et al., 2004) and communication opportunities (Teasley & Rochelle, 1993). A microcomputer-based experiment contains the tools designed to decrease unnecessary laboratory and mental effort that may overload students during data collection and processing (Krajcik & Layman, 1992; Tho & Hussain, 2011); and simulations enable students to construct models by manipulating variables (Kubicek, 2005). Kim and Hannafin (2011) provided a brief but complete description about how technology has been used to assist science learning:

> technology guides the students to focus on critical aspects of problem solving, such as observation of phenomena, identification of evidence, construction of solutions, and collaboration and justification, by taking over tasks that are cognitively less important or completely out of the students' capability (e.g., visualization of complicated scientific theories). (p. 256)

They pointed out how science teaching can be successful if technology is meaningfully and properly used to assist the development of students' inquiry ability. Therefore, ICTs are indispensable in science teaching nowadays from the perspectives of science content representations and scientific practices. Knowledge of curriculum planning and designing with ICTs deserves science teachers' extra attention to develop and initiate their TPACK-P.

2.3.3 Knowledge of Classroom Instruction

Teaching practice can be viewed as the performance stage or construction site where teachers' understanding of students' conditions is a prerequisite and curricula are customized accordingly. Teachers construct their knowledge of subject matter and pedagogy when they are in college or university, but these isolated domains of knowledge are vague or inactive when faced with pedagogical concerns in real classroom teaching practices (Gess-Newsome & Lederman, 1993). It is teaching practice that coherently weaves these isolated domains of knowledge into a set of craft knowledge that is applied to specific topics and to specific students. Shulman (1987) called such a set of teaching knowledge as "wisdom of practice" (p. 9) that accumulates with increased teaching experience and teachers' reflections on their actions.

Both applying ICT to instructional management (Dimension F) and infusing ICT into teaching contexts (Dimension G) were rated important to TPACK-P. Previous research findings have indicated the importance of teachers developing strategies to cope with situations in classrooms such as contingency or management (Atjonen, Korkeakoski, & Mehtalainen, 2011; Garrahy, Kulinna, & Cothran, 2005), which is also reflected in the expert teachers' ratings in this study. However, the findings regarding comparatively lower importance ratings on ICT-assisted instructional management echoed the scant attention that teachers paid to classroom management (Borko & Putnam, 1996). In the past decade, educators and educational software designers have been working on developing classroom-management systems for lowering the burden of management duties or other miscellaneous issues. Learning management systems (LMS), such as WebCT and Blackboard, offer an e-learning environment where students' learning needs can be adaptively accommodated with functions of asynchronous and synchronous communications, content development and delivery, formative and summative assessment, and class and user management (Coates, James, & Baldwin, 2005). Current LMS systems are not perfect, but they can be improved by personally adapting them to students' learning needs (Despotović-Zrakić, Marković, Bogdanović, Barać, & Krčo, 2012). Quality teaching and learning take place once classroom routines are well established and maintained and students' activity performances are in line with learning objectives (Kounin, 1970; Rink, 2002). However, since most teachers tend to spend time working on their subject content instruction rather than classroom management, it is necessary to increase teachers' knowledge of LMS on one hand and improve the platforms to be more user-friendly and intuitive on the other.

2.4   Final Remarks

Teachers' knowledge is further developed and consolidated through the accumulation of teaching practices. Cochran-Smith and Lytle (1999) pointed out that there should be three levels of teacher knowledge: knowledge *for* practice, knowledge *in* practice, and knowledge *of* practice. Knowledge gained from teacher education programs or textbooks would not produce quality, knowledgeable teachers; on the contrary, the knowledge that teachers gain from teaching practices and reflections in and after teaching refines and enriches what they acquired previously. The longitudinal transformation and infusion of these three types of knowledge help develop the teachers' professional knowledge.

Many researchers have endeavored to define the composition of TPACK in order to evaluate the quality of teacher education. The way they analyze teachers' TPACK is similar to the way cooking schools qualify chefs based on their culinary knowledge of food, cooking utensils, cooking strategies, and some hybrid knowledge. This chapter attempts to approach teachers' TPACK from a more practical perspective that is similar to investigating chefs' culinary knowledge according to how they demonstrate food preparation to the point of presenting and serving the food that satisfies their customers' different dietary needs. Knowing learners' starting points could be viewed as the prerequisite for designing curricula to accommodate individual or certain groups of students' learning needs. Having good knowledge about whom and what to be taught can help teachers design and deliver effective instruction with more emphases on students' learning progress – no learning, no teaching.

Inherent in the fundamental concept of TPACK is a unique set of teacher knowledge with which instruction is a pedagogical product after consideration of learners, pedagogy, content, and technological affordances. The framework of TPACK-P identifies eight dimensions of teachers' knowledge when teaching science with ICT. Though all dimensions were rated as important, the knowledge necessary for making content instruction more comprehensible received the two highest ratings (i.e., using ICT to understand subject content and using ICT representations to present instructional representations). This suggests that science teachers value students' subject content learning more than the needs to evaluate students' learning and to seek better classroom management. On the contrary, the lower importance rating for using technology to assist classroom management suggests that the potential of ICT in this area deserves science educators' consideration in teacher education programs. With the dimensions and indicators identified in TPACK-P, we would like to point out the knowledge teachers need to rely on, especially in the case of science teachers, when designing and carrying out their instruction. Although this chapter outlines the knowledge that expert teachers develop from their lengthy teaching experiences with technology, more studies need to investigate how much experts really know and practice their teaching. Such comparisons would help us to know where to refine our teacher education programs toward practical directions.

References

Ainsworth, S. (2006). DeFT: A conceptual framework for considering learning with multiple representations. *Learning and Instruction, 16*(3), 183–198.

Angeli, C., & Valanides, N. (2009). Epistemological and methodological issues for the conceptualization, development, and assessment of ICT-TPCK: Advances in technological pedagogical content knowledge (TPCK). *Computers & Education, 52*(1), 154–168.

Archambault, L. M., & Barnett, J. H. (2010). Revisiting technological pedagogical content knowledge: Exploring the TPACK framework. *Computers & Education, 55*(4), 1656–1662.

Archambault, L. M., & Crippen, K. (2009). Examining TPACK among K–12 online distance educators in the United States. *Contemporary Issues in Technology and Teacher Education, 9*, 71–88.

Atjonen, P., Korkeakoski, E., & Mehtalainen, J. (2011). Key pedagogical principles and their major obstacles as perceived by comprehensive school teachers. *Teachers and Teaching, 17*(3), 273–288.

Borko, H., & Putnam, R. T. (1996). Learning to teach. In D. C. Berliner & R. C. Calfee (Eds.), *Handbook of educational psychology* (pp. 673–708). New York: Macmillan.

Coates, H., James, R., & Baldwin, G. (2005). A critical examination of the effects of learning management systems on university teaching and learning. *Tertiary Education and Management, 11*, 19–36.

Cochran, S. W. (1983). The Delphi method: Formulating and refining group judgements. *Journal of the Human Sciences, 2*(2), 111–117.

Cochran, K. F., DeRuiter, J. A., & King, R. A. (1993). Pedagogical content knowing: An integrative model for teacher preparation. *Journal of Teacher Education, 44*(4), 263–272.

Cochran-Smith, M., & Lytle, S. L. (1999). Relationships of knowledge and practice: Teacher learning in communities. *Review of Research in Education, 24*, 249–305.

Cox, S. (2008). *A conceptual analysis of technological pedagogical content knowledge.* Unpublished doctoral dissertation, Brigham Young University, Provo, Utah.

Davis, E. A., & Krajcik, J. (2005). Designing educative curriculum materials to promote teacher learning. *Educational Researcher, 34*(3), 3–14.

De Jong, O., van Driel, J. H., & Verloop, N. (2005). Preservice teachers' pedagogical content knowledge of using particle models in teaching chemistry. *Journal of Research in Science Teaching, 42*(8), 947–964.

Delbecq, A. L., van de Ven, A. H., & Gustafson, D. H. (1975). *Group techniques for program planning.* Glenview, IL: Scott Foresman.

Despotović-Zrakić, M., Marković, A., Bogdanović, Z., Barać, D., & Krčo, S. (2012). Providing adaptivity in Moodle LMS courses. *Educational Technology & Society, 15*(1), 326–338.

Duschl, R. A., & Wright, E. (1989). A case study of high school teachers' decision making models for planning and teaching science. *Journal of Research in Science Teaching, 26*(6), 467–501.

Feiman-Nemser, S., & Remillard, J. (1996). Perspectives on learning to teach. In F. B. Murray (Ed.), *The teacher educator's handbook: Building a knowledge base for the preparation of teachers* (pp. 63–91). San Francisco: Jossey-Bass.

Garrahy, D. A., Kulinna, P. H., & Cothran, D. J. (2005). Voices from the trenches: An exploration of teachers' management knowledge. *Journal of Educational Research, 99*(1), 56–63.

Gess-Newsome, J. (2002). Pedagogical content knowledge: An introduction and orientation. In J. Gess-Newsome & N. G. Lederman (Eds.), *PCK and science education* (pp. 3–17). New York: Kluwer.

Gess-Newsome, J., & Lederman, N. G. (1993). Preservice biology teachers' knowledge structures as a function of professional teacher education: A year-long assessment. *Science Education, 77*(1), 25–45.

Graham, C. R. (2011). Theoretical considerations for understanding technological pedagogical content knowledge (TPACK). *Computers & Education, 57*(3), 1953–1960.

Harris, J., & Hofer, M. (2009). Instructional planning activity types as vehicles for curriculum based TPACK development. In C. D. Maddux (Ed.), *Research highlights in technology and teacher education 2009* (pp. 99–108). Chesapeake, VA: AACE.

Hasson, F., Keeney, S., & McKenna, H. (2000). Research guidelines for the Delphi survey technique. *Journal of Advanced Nursing, 32*(4), 1008–1015.

Jones, A., & Moreland, J. (2005). The importance of pedagogical content knowledge in assessment for learning practices: A case study of a whole school approach. *Curriculum Journal, 16*(2), 193–206.

Kim, M. C., & Hannafin, M. J. (2011). Scaffolding 6th graders' problem solving in technology-enhanced science classrooms: A qualitative case study. *Instructional Science, 39*(3), 255–282.

Koehler, M. J., & Mishra, P. (2005). What happens when teachers design educational technology? The development of technological pedagogical content knowledge. *Journal of Educational Computing Research, 32*(2), 131–152.

Kounin, J. S. (1970). *Discipline and group management in classrooms.* New York: Holt, Rinehart & Winston.

Kozma, R. (2003). The material features of multiple representations and their cognitive and social affordances for science understanding. *Learning and Instruction, 13*(2), 205–226.

Krajcik, J. S., & Layman, J. W. (1992). Microcomputer-based laboratories in the science classroom. In F. Lawrenz, K. Cochran, J. Krajcik, & P. Simpson (Eds.), *Research matters to the science teacher*. Manhattan, KS: National Association of Research in Science Teaching.

Kubicek, J. P. (2005). Inquiry-based learning, the nature of science, and computer technology: New possibilities in science education. *Canadian Journal of Learning and Technology, 31*(1), Winter. Retrieved from http://www.cjlt.ca/index.php/cjlt/article/view/149/142

Magnusson, S., Krajcik, J., & Borko, H. (1999). Nature, sources, and development of pedagogical content knowledge for science teaching. In J. Gess-Newsome & N. G. Lederman (Eds.), *Examining pedagogical content knowledge: The construct and its implications for science education* (pp. 95–132). Dordrecht, The Netherlands: Kluwer.

Mayer, R. E. (1999). *The promise of educational psychology: Learning in the content areas*. Upper Saddle River, NJ: Prentice Hall.

Mayer, R. E. (2009). *Multimedia learning* (2nd ed.). New York: Cambridge University Press.

McEwan, H., & Bull, B. (1991). The pedagogic nature of subject knowledge. *American Educational Research Journal, 28*(2), 316–334.

McNair, S. (2004). "A" is for assessment. *Science and Children, 42*(1), 18–21.

Mishra, P., & Koehler, M. J. (2006). Technological pedagogical content knowledge: A framework for teacher knowledge. *Teachers College Record, 108*(6), 1017–1054.

National Research Council. (2012). In H. Quinn, H. A. Schweingruber, & T. Keller (Eds.), *A framework for K–12 science education: Practices, crosscutting concepts, and core ideas*. Washington, DC: National Academies Press.

Osborne, J., Collins, S., Ratcliffe, M., Millar, R., & Duschl, R. (2003). What "ideas-about science" should be taught in school science? A Delphi study of the expert community. *Journal of Research in Science Teaching, 40*(7), 692–720.

Otero, V. K. (2006). Moving beyond the 'get it or don't' conceptions of formative assessment. *Journal of Teacher Education, 57*(3), 247–255.

Quintana, C., Reiser, B. J., Davis, E. A., Krajcik, J., Fretz, E., Duncan, R. G., et al. (2004). A scaffolding design framework for software to support science inquiry. *Journal of the Learning Sciences, 13*(3), 337–386.

Rink, J. E. (2002). *Teaching physical education for learning*. Boston: McGraw-Hill.

Segall, A. (2004). Revisiting pedagogical content knowledge: The pedagogy of content/The content of pedagogy. *Teaching and Teacher Education, 20*(5), 489–504.

Shavelson, R. J., & Stern, P. (1981). Research on teachers' pedagogical judgments, decisions, and behavior. *Review of Educational Research, 51*, 455–498.

Shulman, L. S. (1986). Those who understand: Knowledge growth in teaching. *Educational Researcher, 15*(2), 4–14.

Shulman, L. S. (1987). Knowledge and teaching: Foundations of the new reform. *Harvard Educational Review, 57*(1), 1–22.

Teasley, S. D., & Rochelle, J. (1993). Constructing a joint problem space: The computer as a tool for sharing knowledge. In S. P. Lajoie & S. J. Derry (Eds.), *Computers as cognitive tools* (pp. 229–258). Hillsdale, NJ: Lawrence Erlbaum.

Tho, S. W., & Hussain, B. (2011). The development of a microcomputer-based laboratory (MBL) system for gas pressure law experiment via open source software. *International Journal of Education and Development using Information and Communication Technology, 7*(1), 42–55.

Thompson, J., Christensen, W., & Wittmann, M. (2011). Preparing future teachers to anticipate student difficulties in physics in a graduate-level course in physics, pedagogy, and education research. *Physical Review Special Topics–Physics Education Research, 7*(1). doi:10.1103/PhysRevSTPER.7.010108.

Tigelaar, D. E. H., Dolmans, D. H. J. M., Wolfhagen, I. H. A. P., & van der Vleuten, C. P. M. (2004). The development and validation of a framework for teaching competencies in higher education. *Higher Education, 48*(2), 253–268.

Treagust, D., Chittleborough, G., & Mamiala, T. (2003). The role of submicroscopic and symbolic representations in chemical explanations. *International Journal of Science Education, 25*(11), 1353–1368.

Turoff, M. (1970). The design of a policy Delphi. *Technological Forecasting and Social Change,* *2*(2), 149–171.

van der Meji, J., & de Jong, T. (2006). Supporting students' learning with multiple representations in a dynamic simulation-based learning environment. *Learning and Instruction, 16*(3), 199–212.

van Driel, J. H., Beijaard, D., & Verloop, N. (2001). Professional development and reform in science education: The role of teachers' practical knowledge. *Journal of Research in Science Teaching, 38*(2), 137–158.

van Driel, J. H., Verloop, N., & de Vos, W. (1998). Developing science teachers' pedagogical content knowledge. *Journal of Research in Science Teaching, 35*(6), 673–695.

Verdi, M. P., Johnson, J. T., Stock, W. A., Kulhavy, R. W., & Whitman-Ahern, P. (1997). Organized spatial displays and texts: Effects of presentation order and display type on learning outcomes. *Journal of Experimental Education, 65*(4), 303–317.

Wu, H.-K., Krajcik, J., & Soloway, E. (2001). Promoting understanding of chemical representations: Students' use of a visualization tool in the classroom. *Journal of Research in Science Teaching, 38*(7), 821–842.

Wu, H.-K., Lin, Y.-F., & Hsu, Y.-S. (2013). Effects of representation sequences and spatial ability on students' scientific understandings about the mechanism of breathing. *Instructional Science, 41*(3), 555–573.

Yeh, Y.-F., Hsu, Y.-S., Wu, H.-K., Hwang, F.-K., & Lin, T.-C. (2014). Developing and validating technological pedagogical content knowledge-practical (TPACK-practical) through the Delphi survey technique. *British Journal of Educational Technology, 45*(4), 707–722. doi:10.1111/bjet.12078.

Young, W. H., & Hogben, D. (1978). An experimental study of the Delphi technique. *Education Research Perspective, 5*, 57–62.

Chapter 3
The Current Status of Science Teachers' TPACK in Taiwan from Interview Data

Tzu-Chiang Lin and Ying-Shao Hsu

Teachers' knowledge about technology-infused instruction has recently attracted much research attention. This chapter focuses on science teachers' technological pedagogical and content knowledge (TPACK) in the practical context of teaching, namely, TPACK-Practical (TPACK-P). The proposed framework of TPACK-P includes three major domains—assessments, planning and designing, and teaching practice—that are theoretically transformed from the perspectives of pedagogical content knowledge (PCK). To explore science teachers' TPACK-P, 40 in-service teachers were interviewed, and a coding scheme was developed to analyze the interview responses. The findings indicated that the science teachers generally know how to adopt technologies in teaching within each domain of TPACK-P. A cluster analysis based on the participants' level of TPACK-P categorized their patterns of knowledge. Three groups of science teachers emerged from these analysis categories: infusive application, transition, and plan and design emphasis. The infusive application group represents science teachers with sophisticated levels of TPACK-P across the three domains; the transition group includes science teachers whose knowledge achieved average levels across the three dimensions. However, the plan and design emphasis group refers to the science teachers who were more knowledgeable about planning and designing technology-infused teaching than about the assessment and teaching practice domains. The overall results indicate that the knowledge of planning and designing may be a more independent part in TPACK-P that supports science teachers' implementation of technology-infused teaching. The revealed patterns of these science teachers' TPACK-P may provide the groundwork

T.-C. Lin
Graduate Institute of Digital Learning and Education, National Taiwan University of Science and Technology, Taipei, Taiwan

Y.-S. Hsu (✉)
Graduate Institute of Science Education, National Taiwan Normal University, Taipei, Taiwan
e-mail: yshsu@ntnu.edu.tw

© Springer Science+Business Media Singapore 2015
Y.-S. Hsu (ed.), *Development of Science Teachers' TPACK*,
DOI 10.1007/978-981-287-441-2_3

for developing instruments to evaluate science teachers' competence in teaching with technologies.

3.1 Introduction

In past decades, educational reforms involving information and communication technologies (ICTs) have changed the context of science classrooms worldwide (Lee et al., 2011; Linn, 2003). ICTs have modernized knowledge communication in science education and expanded learning approaches such as collaborative learning (Mäkitalo-Siegl, Kohnle, & Fischer, 2011; Suthers, 2006), inquiry-based learning (Edelson, 2001; Linn, Clark, & Slotta, 2003), project-based learning (ChanLin, 2008; Krajcik, McNeill, & Reiser, 2008), problem solving (Kim & Hannafin, 2011; Serin, 2011), and informal learning environments (Anastopoulou et al., 2012; Ebner, Lienhardt, Rohs, & Meyer, 2010). Regardless of the type or amount of technology applied in classrooms, teachers are still the key to facilitate educational reform with ICTs. Several calls about technologies in science teacher education have revealed the need for deeper investigations of teachers' competence to design and conduct effective technology-enhanced instruction (Angeli & Valanides, 2005, 2009; Lin, Tsai, Chai, & Lee, 2013). Moreover, much of effective technology-enhanced instruction involves practical knowledge regarding how a teacher makes sense or establishes application of ICTs in classrooms. This chapter focuses on the current state of Taiwanese science teachers' competence with ICT from a practical perspective.

3.1.1 The Role of TPACK

Teacher educators and policy makers have tried to establish norms for teachers' knowledge about effective teaching and classroom practices (Ball, Thames, & Phelps, 2008). In recent years, technological pedagogical and content knowledge (TPACK) has been addressed to portray teachers' competence to teach in technology-infused environments (Lin et al., 2013; Voogt, Fisser, Pareja Roblin, Tondeur, & van Braak, 2013). For example, Mishra and Koehler (2006) used notions of pedagogical content knowledge (PCK; Shulman, 1986, 1987) to develop an integrative model illustrating the intersections of content knowledge (CK), pedagogical knowledge (PK), and technological knowledge (TK) to represent teachers' knowledge about discipline-specific teaching with ICTs. Researchers have also tried to identify and to measure teachers' TPACK using varied methods such as questionnaires (Archambault & Crippen, 2009; Lee & Tsai, 2010; Schmidt et al., 2009), tests (Angeli & Valanides, 2009; Kramarski & Michalsky, 2010), and interpretative interviews (Jimoyiannis, 2010; Niess, 2005). These results have established TPACK as a trustworthy construct and the basis for evaluation of teacher professional development.

Theoretically, TPACK refers to the knowledge about teaching academic contents of a specific discipline with ICTs. Therefore, science teachers' TPACK may be divergent in nature, which will be apparent, while teachers plan, enact, and evaluate lessons in different subject domains and classrooms. Previous empirical studies have found that science teachers' perceived TPACK was distinct from teachers with dissimilar academic expertise (Lin et al., 2013). Current TPACK-related research on science teacher education has mainly employed Mishra and Koehler's (2006) model of TPACK and has investigated internal components of the knowledge system (Jimoyiannis, 2010). However, it is necessary to investigate teachers' knowledge from a practical context situated in science classrooms to document how the PCK is transformed into TPACK (Angeli & Valanides, 2009; Graham, 2011). From this emerges the need to establish a comprehensive foundation for improving contemporary science teacher education based on the TPACK rationale.

3.1.2   TPACK in Science Teacher Education

Science teacher education has emphasized the need to apply general pedagogical ideas to the specific demands and contexts of learning science at different school levels. Most science teacher education programs assume that preservice teachers on one hand acquire their science content knowledge from coursework in the academic science department. On the other hand, they develop general pedagogical knowledge from education and educational psychology coursework. Such teacher education models assume that science curricula and instruction coursework as well as clinical experiences will help preservice teachers integrate their academic science and general PK into discipline-specific PK. This knowledge, called PCK, is deemed a crucial part of teachers' competence in successful science teaching. How well science teachers integrate science content and their teaching experiences into their PCK has been questioned because the theory–practice gap continues to exist. The conversion of theoretical knowledge into teaching practices appears to be a career-long process or struggle for teachers, which involves transforming as well as integrating CK and PK into PCK.

Science educators have expressed a consensus that ICTs bring great impacts to learning and teaching in science, but merely emphasizing either computer skills or pedagogy in teacher education has little benefit in preparing teachers to adequately and effectively utilize technology in their careers (Hughes, 2005; Keating & Evans, 2001; Parkinson, 1998). Like *the song remains the same* addressed by Mishra, Koehler, and Kereluik (2009), teacher educators still seek ways to enhance teachers' capability with technology. Consequently, a burgeoning consensus of teacher knowledge about teaching with technology, namely, TPACK, has more recently inspired teacher educators and researchers (Koehler & Mishra, 2005; Niederhauser & Stoddart, 2001; Web & Cox, 2004). TPACK—formerly the acronym TPCK (Thompson & Mishra, 2008)—provides a valuable framework on which to determine whether a teacher is able to effectively design and conduct technology-infused instruction (Angeli & Valanides, 2005; Mishra & Koehler, 2006). Angeli and Valanides (2009) suggested that teachers with sufficient TPACK may gradually understand the specifics of technological tools with regard to the relationships among technological tools, instructional designs, contents, student characteristics, and teaching contexts.

TPACK has attracted teacher educators' attention and focus on the issues associated with teachers' utilization of ICT in classrooms. Searching academic databases indicates that TPACK was, and still is, a hot topic in the field of educational technology and teacher education (Chai, Koh, & Tsai, 2010; Voogt et al., 2013). TPACK is deemed as having the potential to recognize and predict how teacher educators' interventions affect teachers' competence from a knowledge perspective (Graham, 2011). Moreover, a successful teacher with sufficient TPACK may be able to develop proper strategies and representations to accomplish fruitful teaching with technology.

Teachers' practical knowledge refers to how well teachers understand and apply their professional activities in the teaching context (van Driel, Beijaard, & Verloop, 2001). A similar theory–practice gap exists for TPACK: the knowledge that is directly associated with teachers' practical experience of teaching with ICTs (Graham, 2011). The integrated model of TPACK seems insufficient to explain the process of how teachers build knowledge about using ICT in science teaching contexts (Angeli & Valanides, 2009). Intrinsic influence from teaching context to TPACK is still vague in the model (Angeli & Valanides, 2009). Therefore, we adopt an extended model of TPACK-P (Yeh, Hsu, Wu, Hwang, & Lin, 2014) to unveil the struggle and status of science teachers' practical knowledge in a technology-infused teaching context and to clarify specific features of TPACK-P regarding the domains of assessment, planning and designing, and teaching practices.

This chapter draws on the recently proposed framework of TPACK-P (Yeh et al., 2014) to document science teachers' knowledge about teaching with technologies. In order to concisely identify the patterns of such an extended form of TPACK from in-service science teachers, we reorganized the original framework into three major domains of TPACK. Hence, we endeavor in this study to clarify science teachers' TPACK-P from how they know about (a) conducting assessment with ICTs, (b) planning and designing teaching with ICTs, and (c) processing practical teaching activities with ICTs. Teachers with well-developed TPACK-P are likely to make effective use of technologies in knowing their students with assessments (Jang & Tsai, 2012), in presenting contents with pertinent planning and design (Lundeberg, Bergland, Klyczek, & Hoffman, 2003; Niess, 2005), and in dealing with classroom management (Graham, 2011; Koehler, Mishra, & Yahya, 2007). Hsu, Wu, and Huang (2007) adopted Sandholtz, Ringstaff, and Dwyer's (1997) suggestion to classify science teachers into five stages (i.e., entry, adoption, adaption, appropriation, and invention) when utilizing technological tools in instruction. Although their survey results showed a hierarchy of science teachers' professional activities with ICTs, there is still a lack of evidence to reveal the features of teachers' TPACK knowledge within these five stages. Hence, this chapter reports science teachers' TPACK-P in terms of their authentic teaching experiences based on interview data.

3.2   Revealing Science Teachers' TPACK-P

In order to reveal science teachers' TPACK-P, we adopted the rationale that science teachers' PCK is transformed during its application in a technology-infused context into TPACK-P. First, we explored science teachers' knowledge about conducting assessments with ICTs, that is, using ICTs to know more about students, identify students' learning difficulties, assist different characteristics of learners, know the types of technology-infused assessment approaches, identify the differences between technology-infused assessments and traditional assessments, and utilize e-assessments for detecting students' learning progress. Second, we identified science teachers' TPACK-P about instructional planning and designing by investigating their ICT uses to better understand subject contents, identify the topics that can be better presented with ICTs, use appropriate ICT representations to present instructional contents, and apply appropriate teaching strategies in ICT-infused instructions. Third, we investigated science teachers' TPACK-P about teaching practices with regard to their use of ICTs to indicate differences between traditional and ICT-infused instruction, indicate the influences of different ICT instructions, indicate substitute plans for technology-infused instruction, and facilitate instructional management. In summary, this chapter presents how we investigated science teachers' TPACK-P in Taiwan.

3.2.1   Methods to Reveal Science Teachers' TPACK-P

A mixed methods approach (Creswell, 2008; Creswell & Plano Clark, 2011) was applied to explore and interpret the participating science teachers' TPACK-P regarding the three domains: assessment, planning and designing, and teaching practice. Furthermore, we categorized science teachers based on their TPACK-P using a cluster analysis technique.

The current investigation recruited participants with different academic majors, teaching experience, and experience of winning educational awards for technology-infused instruction. Forty in-service science teachers in northern Taiwan were purposefully selected. The authors acquired each science teachers' permission to participate through private invitations by a telephone call or email. Although these science teachers' experiences of teaching with ICTs varied, all of them had participated in professional development programs focused on technologies in science instruction. Furthermore, we invited only those teachers who had taught science in high school for more than 5 years to ensure that they had enough experience in teaching science with ICTs.

Semistructured interviews were employed as the major approach to collect data about these science teachers' TPACK-P. A group of science educators (three professors, one postdoctoral researcher, and two doctoral students) developed the interview protocol through panel meetings and went through several roundtable discussions to ensure that the questions were appropriate to probe for science teachers'

TPACK-P. The interviews were first administered to several science teachers as pilot trials to validate the interview process (Guba & Lincoln, 1989) and to eliminate inappropriate questions. The interview questions are provided in the Appendix of this chapter.

The semistructured interviews were conducted by a postdoctoral researcher and two doctoral students; each interviewer was familiar with semistructured interview techniques. All participants agreed with audiotaping the interview. The interviews took 40–60 min. It is worth noting that the interviewers tried to avoid yes/no responses by asking follow-up elaboration and clarification questions unless the participant indeed had no idea about the question.

3.2.2 Analysis of Science Teachers' TPACK-P

The audiotaped interviews were transcribed as verbatim texts. Transcriptions with ambiguity were returned to interviewees for verification and clarification. Thereafter, we analyzed the interview transcriptions simultaneously with thematic coding (Flick, 2002) and axial coding (Strauss & Corbin, 1990) approaches. First, all responses were aggregated to comprehensively summarize all features of science teachers' TPACK-P and then classified into thematic categories in accordance with Sandholtz et al.'s (1997) five stages of practical teaching with technologies (i.e., entry, adoption, adaption, appropriation, and invention). However, we encountered difficulty in fitting part of the TPACK-P features into the preliminary categories; for example, in the entry stage, "lack of use of technology in teaching" was not clearly differentiated from "no idea of technology application."

Therefore, we applied constant comparative methods to reinspect the levels of TPACK-P. We performed axial coding repeatedly to reveal similarities and discrepancies in interview narratives and, thus, refine the categorization. After several discussions to specify the levels, the final thematic coding categories were reformed into five categories for assessment, planning and designing, and teaching practice (Table 3.1). The categories were defined as:

- 0—No idea—represents teachers without any notion of technological application in teaching; for example, they are not conscious of using an audience response system (Kay & LeSage, 2009) to diagnose students' learning
- 1—Lack of use—represents situations that teachers simply expressed their understanding of ICTs for instruction (e.g., computer-supported, collaborative learning environment) but did not make use of it in their classes
- 2—Simple adoption—represents teachers' ICT usage in teaching without the statements related to the purpose, employment, or effect of applying ICTs
- 3—Infusive application—represents teachers' successful integration of ICTs in teaching while they clearly describe the purpose, employment, and effect of their integration

Table 3.1 Coding categories of TPACK-P and examples of category components

Domain	Assessment	Planning and designing	Teaching practice
Level 0—no idea	Never consider the possibility of using technologies in assessment	Never consider any instructional goal for technology-infused teaching	Never consider the difference between technology-infused and traditional teaching practice
	Have no idea about technological tools for assessment such as audience response system	Have no idea about proper technological representation applied in teaching	Have no idea about the effects that different technological tools can bring to teaching practice
	Not able to differentiate technology-infused and traditional assessment	Have no idea about teaching strategy for technology-infused teaching	Have no idea about applying technology to support instructional management
Level 1—lack of use	Indicate the lack of useful hardware or software to conduct assessment	Indicate that time consuming in course preparation leads to the lack of using technologies in planning and designing teaching	Indicate the avoidance of using ICTs in teaching practice based on various reasons
	Indicate that technologies are main supplement of classroom lecture rather than useful tools for assessment	Indicate that traditional lecture is much more important than technology-infused teaching	Indicate the application of technologies in instructional management gives rise to extra teaching loading
	Indicate that technology-infused assessment may be unfair	Indicate no need for making use of specific strategy for technology-infused teaching	Indicate the worry about risks that derive from practical teaching with technologies
Level 2—simple adoption	Simply mention the application of technologies to know students' comprehension of science phenomena	Simply mention why some learning units are suitable for using technology-infused teaching	Simply mention how technological tools impress students in teaching practice
	Simply mention the usage of online assessment, online questionnaire, or audience response system	Simply mention the limitation in planning and design that may be influenced by physical environment of classroom or students' responses	Simply mention the comparison of instructional aids in technology-infused and traditional teaching practice
	Simply mention the characteristics of technology-infused assessment	Simply mention useful teaching strategies (not specified) for enhancing students' motivation, engagement, and concentration	Simply mention the usage of software in instructional management

(continued)

Table 3.1 (continued)

Domain	Assessment	Planning and designing	Teaching practice
Level 3—infusive application	Describe the process of applying technological tools in conducting assessment	Describe the process of getting better understanding of academic content and teaching strategies by using technologies	Describe how technology-infused teaching especially supports learning of students who get lower achievement in traditional teaching context
	Describe in detail the process of using online assessment, online questionnaire, and audience response system to know students' ideas	Describe the integration of technologies in teaching planning and design to support students' scientific processes such as data collection, analysis, and presentation	Describe what effects that different technological tools may bring to teaching practice
	Describe the process of using technologies to know students' difficulties or alternative conceptions in learning science	Describe in detail about how to apply a specific strategy, such as computer-supported collaborative learning in planning and designing teaching	Describe in detail about the process of using technological tools to deal with instructional management
Level 4—self-evaluation	Describe how technologies can enhance traditional assessment with a more adaptive and interactive manner	Describe the reflection of successful or fruitless experience about planning and designing technology-infused teaching	Describe the reflection of advantages of technological tools that make changes in traditional teaching practice
	Describe why technology-infused assessment is better for understanding students' alternative conceptions of science	Describe the ability that a teacher shall have to efficiently integrate technologies in planning and designing technology-infused teaching	Describe multiple substitutions for possible contingency in technology-infused teaching practice
	Describe one's own experience of using technology-infused assessment and hence self-evaluate the relevant competence and possible improvement of teaching	Describe the adaptive adjustment of strategies to fit in with instructional goal of technology-infused teaching	Describe the necessity to self-evaluate application of technological tools in instructional management

- 4—Self-evaluation—represents teachers who expressed their knowledge of examining and regulating their teaching with ICTs (e.g., evaluating the design of technology-infused approaches compared to conventional teaching) to meet students' needs

The final coding themes and categorized features of TPACK-P summarized in Table 3.1 were utilized to code participants' responses to all interview questions. Since the responses might be partially categorized into different levels of TPACK-P, we deemed that the higher level might present a more sophisticated view of the teacher's TPACK-P for that question. Therefore, we assigned an achieved level to the responses for each interview question. The overall agreement of two coders achieved 0.96 and indicated a congruent coding process. Subsequently, we performed cluster analysis (Lorr, 1983) with hierarchical clustering technique on the labeled interview responses to group the participants with similar patterns of TPACK-P. In order to illustrate the pattern of these groups of science teachers' TPACK-P, both qualitative and quantitative analyses were conducted. The qualitative findings were based mainly on the interpretations of interview data; the quantitative analyses examined if the groups of teachers showed statistical difference in the three domains of TPACK-P. Due to the concerns for small sample size and the ordinal nature of the data, we applied a nonparametric statistical analysis to identify the difference, namely, a Kruskal–Wallis one-way analysis of variance (ANOVA) with a post hoc Dunn's test.

3.3 Characteristics of Science Teachers' TPACK-P

According to the analyses of descriptive statistics of the coding results on a 0–4 scale, the mean levels of participants' TPACK-P were ($n=40$): overall, $M=2.63$, $SD=0.37$; assessment, $M=2.57$, $SD=0.60$; planning and designing, $M=2.77$, $SD=0.45$; and teaching practice, $M=2.49$, $SD=0.52$. These results implied that the Taiwanese science teachers in this investigation showed an above midrange (2.00 on a 0–4 scale) degree of competence about teaching with ICT integration. At the least, these teachers were capable of stating in general how they adopt technologies in teaching based on their experiences. Inspection of the individual response values

Table 3.2 Descriptive statistics of science teachers grouped by TPACK-P

| | Group | | | | | |
| | 1–Infusive application ($n=18$) | | 2–Transition ($n=10$) | | 3–Plan and design emphasis ($n=12$) | |
Domain of TPACK-P	M	SD	M	SD	M	SD
Assessment	2.94	0.48	2.58	0.50	2.00	0.40
Planning and designing	2.99	0.37	2.41	0.49	2.73	0.32
Teaching practice	2.75	0.48	2.40	0.58	2.17	0.34

indicated variation around the means and performance patterns across the three domains; therefore, further analyses were justified.

The cluster analysis successfully categorized the teachers into three groups in terms of their pattern of coded response levels for all interview questions (Table 3.2). We then interpreted the patterns and described these three groups based on the three domains of science teachers' TPACK-P as follows. First, a group of 18 teachers demonstrated higher and balanced levels in each domain of TPACK-P that approximated level 3—infusive application; we identified this group as "infusive application, IA," for further discussion. Second, the mean level of a group of 10 teachers demonstrated lower levels but balanced performance across the domains of TPACK-P compared to the IA group; their response levels were near to the overall mean value. We identified this group as "transition, TR." Last, a group of 12 teachers showed a thoroughly different pattern of TPACK-P levels: a fairly high mean level in the planning and designing domain and noticeably lower levels in the other two domains than the IA and TR groups. We identified this group as "plan and design emphasis, PD."

The results of Kruskal–Wallis one-way ANOVA indicated that the PD group had a significant ($p<0.001$) main effect across the three TPACK-P domains, but there were no significant main effects for the IA ($p=.18$) and TR ($p=.88$) groups. Based on the results for the PD group, a series of pairwise comparisons were used on the domains using the Dunn's test. The post hoc Dunn's tests revealed that the PD group achieved significantly higher levels in planning and designing than the other two domains (plan and design vs. assessment, $p<.01$; plan and design vs. teaching practice, $p<.05$). These results suggest that these teachers' knowledge about assessment and teaching practice were more similar domains within their TPACK-P and that these two domains are directly related with actual implication of technologies in instruction. Furthermore, the planning and designing competence of these teachers may be a more independent domain and, therefore, have less influence on predicting their teaching implementation. However, such an assertion may need further support and exploration.

Generally, the IA group achieved higher levels because these teachers tended to think about teaching practice with technologies with greater consideration of students' needs. For example, one might consider the possibility to overcome the limits of traditional assessments (i.e., tests with paper and pencil) by applying technologies. Exemplar responses from the interviewees follow:

T23: Tests with paper and pencil can also estimate [students'] affections or something, right? But in fact, this part will be exhausting [on paper and pencil tests]. With technology, not only for [more learning] time, it can present [the assessment] in more [and] different manners. In that way, [students] can [have] his/her own way to answer. For example, someone likes a movie clip, someone likes an animation or something. The animation may replace a large number of words in the test, especially when the wording [in the tests] was [difficult]. In that way, we may make students realize the assessment [can measure] their achievement more precisely.

Some other teachers criticized the overemphasis on rehearsing factual knowledge in traditional assessments. Instead, they highlighted the possibility that technologies can contribute to alternative ways of assessment:

T15: If you really focus on students' development of science literacy, you can never simply aim at their ability to solving questions in an examination. Moreover, students need to develop some practical competence, such as observation of natural phenomena, operating scientific equipment, and so forth. It is not possible to use paper and pencil tests to probe these competencies [of students]. But technological tools such as manipulative simulations in earth science can help you observe the students' operational skills. You can even track their process of thinking by examining the log file recorded in the software.

The student-centered idea was also reflected in how these science teachers designed the technology-embedded teaching. The availability for learning with technology may be first emphasized when planning and designing teaching. The teachers who achieved the highest level (coded as 4—self-evaluation) in the assessment domain also responded with student-centered ideas about designing software that focused on learning:

T23: [In designing teaching,] I will first consider whether this tool is proper or not, as well as if students have corresponding equipment to use in learning, like e-schoolbag [mobile learning equipment] or something. As a result, when the teacher designs some learning software [that is] only available or executable on some platform or browser, this may detract from students' participation. We should consider other tools or interface [to avoid] falling into such situations.

T12: [In designing teaching] what I really care about is the diversity of students, especially from the perspective of motivation and engagement. Designing and integrating a technology-embedded curriculum can help me attract students with lower motivation that possibly resulted from lower cognitive ability. As to students with high academic achievement, the technological tools about science learning, such as some apps for the iPad that they seldom make use of in daily life, may trigger their curiosity and enthusiasm to explore the relevant scientific knowledge.

The IA group teachers clearly defined the manner to manage the interactions with students, such as online communications. The following quotes provide a glimpse of a teacher's idea that related to his knowledge about implementation as well as planning and designing teaching:

T23: We may not apply this [online communication] in normal class. To me, if you want to establish a blog or forum, you must spend time to maintain it. Yes, I am sure such kind of communication is the teachers' responsibility. Furthermore, the administrator [the teacher] must be good at organizing students' statements, discussion and reveal the answers from different viewpoints. For now, I don't think there is a good platform for doing this. Facebook may be a possibility, but for most situations students may just chat, [using a] kind of instant communication [that is] hard to use for learning purposes.

In contrast, the TR group teachers reflected more of a teacher-centered perspective about TPACK-P:

T17: Utilizing technologies in assessment is simply making the test like a game.

T33: Students still need to experience the calculation in tests, somehow just with a more funny way.

T6: I will not imagine that technology is able to provide significant assistance for summative evaluation because this involves the problematic equity of testing with technologies.

These opinions might inhibit them from knowing more about what technology
can do when evaluating both students' achievement and their own instruction.
Moreover, these teachers presented some arbitrary ideas about adopting technologies in instructional planning and designing. Their responses to the interview
questions were, therefore, coded at a lower level (2—simple adoption) of
TPACK-P. For example, one teacher appeared to be subjective about indicating the
factors that might influence technology-embedded instruction:

T17: [The critical factor will] be hardware. This is a quite common problem among us [sci-
ence teachers].
I: Yes, hardware, then what are the other factors?
T17: We must rely on chalk and talk for most situations. Because the planning of such
kind of teaching is more flexible. Even though we know how to use PowerPoint to teach,
but I am just not getting used to [it]. I have no flexibility of control when I am inspired by
something about expanding my teaching designs.

The PD group of teachers presented a different pattern of responses. These teachers tended to be knowledgeable about planning and designing teaching with technology. However, they hold less sophisticated knowledge about applying
technologies in assessment and teaching practice:

T4: I am traditional about assessment because paper and pencil test is the most efficient way
of testing in my [learning] experience.... I am not familiar with the computer classroom in
school, and I am not using any special technological tools in my teaching, even for making
use of the Internet.

On the contrary, she could address why technology was important to satisfy
instructional goals and needs in planning and designing:

T4: When you cannot situate your students in the real-world context, technology can be a
good alternative. For example, you can never have your students experience all kinds of
ecosystems or see all kinds of animals live when you introduce taxonomy. It is also difficult
to help your students understand a complex physiological process such as blood circulation
with verbal explanations. These are the most important reasons that we need computers in
instruction.

These findings indicate that the participating science teachers indeed hold varied
TPACK-P, even though they may be capable of adopting technology in teaching.
The qualities and patterns of TPACK-P can be used to characterize science teachers'
TPACK. These findings provided insights into the development of research tools for
probing teachers' TPACK, while the estimations of TPACK in recent research were
based mostly on self-reported perceptions of such knowledge (Archambault &
Barnett, 2010; Schmidt et al., 2009; Yurdakul et al., 2011). Objective evaluations of
teachers' professional development and knowledge, such as TPACK, may be an
immediate indicator of success or failure of policies administered in teacher education. The findings in this research may inform both teacher educators and stakehold-

ers with a practical direction of developing objective evaluation of science teachers' TPACK.

Still, it is worth noting that many of these teachers met difficulty in responding to some interview questions of the planning and designing domain, even though they presented the highest mean level (2.77) in this part. Almost a quarter (9 of 40) of the participants had no idea about naming and describing a proper strategy to apply technology in planning and designing their teaching. Teacher T25, for example, when asked about describing the strategies she used in instructional design, could only reply that she used technology throughout the teaching process but was unable to indicate the specific strategies applied. She could, however, clearly describe both the topics and technological tools that are suitable for technology-embedded instruction based on her successful classroom experiences. The unfamiliarity with instructional strategies of these participants may imply that science teachers seldom emphasize the educational theories and evidence-based practices applied in their teaching. Instead, their ideas about planning and designing may rely heavily on their teaching or learning experiences.

Previous psychological research has addressed the influence of successful mastery experience over psychometric features about teaching, such as science teaching self-efficacy (Klassen, Tze, Betts, & Gordon, 2010; Lumpe, Czerniak, Haney, & Svetlana, 2012) as well as attitudes and epistemic beliefs (Hofer, 2000; Palmer, 2002). Successful teaching experience is deemed as the most important source contributing to teachers' confidence in accomplishing a specific instructional task, such as teaching with technologies. Furthermore, such experience may affect teachers' beliefs that shape their TPACK-P. These findings suggest that teacher educators should pay more attention to how science teachers acquire mastery experience in future professional development programs regarding their individual characteristics and needs.

3.4 Concluding Remarks

This chapter examined science teachers' practical knowledge about teaching with technology through interviewing different science-subject teachers. Based mainly on their responses, we concluded the construct of TPACK-P acts as a serviceable framework that recognizes science teachers' knowledge about applying technologies in respect of assessment, planning and designing, and teaching practices. The current status of these Taiwanese science teachers' TPACK-P revealed a triad that indicated they presented an unbalanced combination of the three domains of TPACK.

Moreover, the findings provide preliminary value for future development of assessment tools that are reliable for evaluating science teachers' TPACK-P. This may also provide insights for estimating the effect of professional learning on both in-service and preservice science teacher education. In order to establish the standard to assess science teachers' TPACK-P, there is still a need for more comprehensive evidence sources. Research findings from different social or cultural contexts, approaches other than qualitative settings, as well as participants with diverse teaching experience (e.g., preservice teachers) may contribute to understanding teachers' overall TPACK and the patterns of their TPACK-P.

Knowledge of instructional planning and designing may be partly independent of knowledge about implementation of instruction with regard to assessment and teaching practices. This implies a direction for future investigations to reveal science teachers' other characteristics that may affect their TPACK-P, such as beliefs about science teaching (Lumpe et al., 2012) and conceptions of science teaching (Yung, Zhu, Wong, Cheng, & Lo, 2013). The findings provided in this chapter suggest that science teacher educators and policy makers should conduct programs of improving science teachers' TPACK-P based on technology affordance that address the needs of teaching and learning within the current educational milieu.

Appendix: Interview Questions

Assessment Domain

1. How does technology help you realize students' individual differences?
2. How does technology help you realize students' characteristics of learning?
3. How does technology help you recognize students' difficulties about learning?
4. Can you provide some examples of using proper technological tools to afford different students' learning?
5. Is there any other way that can help you make use of adaptive technologies to assist students' learning?
6. Do you know technology-infused assessment?
7. Have you ever used technological tools to conduct assessment?
8. Can you design technological tools (including hardware and software) for assessment?
9. Can you recognize the difference between technology-infused assessment and traditional assessment?
10. How do technologies help you in formative assessment? How do technologies help you in summative assessment?

Planning and Designing Domain

1. How do you apply technologies to improve your understanding about academic content in teaching?
2. Is technology-infused teaching especially suitable for some academic content in teaching? Why?

3. What factors will affect your technology-infused teaching when you try to conduct planning and designing such teaching? How do you deal with these possible factors?
4. What goal do you have when you conduct planning and designing technology-infused teaching? How do you follow the goal?
5. Have you ever collected teaching materials by using ICTs? Can you provide some examples?
6. Will you prepare any substitutes when you conduct planning and designing technology-infused teaching? Can you provide some examples?
7. In planning and designing technology-infused teaching, how do you choose suitable technological tools to present your teaching? What about this in varied learning context such as normal classroom and laboratory? Can you provide some examples?
8. In planning and designing technology-infused teaching, is there any suitable teaching method or teaching strategy? Why?
9. What will you expect about your students' responses while you apply suitable teaching method or teaching strategy in technology-infused teaching?

Teaching Practice Domain

1. In your teaching experience, how does technology affect your course proceeding? Is there any difference when there is no technology infused?
2. In your teaching experience, how does technology affect your students' learning performance? Is there any difference when there is no technology infused?
3. In your teaching experience, how does technology affect your students' motivation? Is there any difference when there is no technology infused?
4. In your teaching experience, have you ever applied technological tools with varied characteristics to support course proceeding? How do these tools affect your teaching?
5. In technology-infused teaching, how do you deal with the contingency of hardware and software? What will you do if the contingency delays your teaching schedule?
6. Do you know any technological tools for instructional management? Can you provide some examples based on your teaching experience?
7. What is the advantage of applying technologies in instructional management? What about disadvantages?

References

Anastopoulou, S., Sharples, M., Ainsworth, S., Crook, C., O'Malley, C., & Wright, M. (2012). Creating personal meaning through technology-supported science inquiry learning across formal and informal settings. *International Journal of Science Education, 34*(2), 251–273.

Angeli, C., & Valanides, N. (2005). Preservice elementary teachers as information and communication technology designers: An instructional systems design model based on an expanded view of pedagogical content knowledge. *Journal of Computer Assisted Learning, 21*(4), 292–302.

Angeli, C., & Valanides, N. (2009). Epistemological and methodological issues for the conceptualization, development, and assessment of ICT–TPCK: Advances in technological pedagogical content knowledge (TPCK). *Computers & Education, 52*(1), 154–168.

Archambault, L. M., & Barnett, J. H. (2010). Revisiting technological pedagogical content knowledge: Exploring the TPACK framework. *Computers & Education, 55*(4), 1656–1662.

Archambault, L. M., & Crippen, K. (2009). Examining TPACK among K-12 online distance educators in the United States. *Contemporary Issues in Technology and Teacher Education, 9*(1), 71–88.

Ball, D. L., Thames, M. H., & Phelps, G. (2008). Content knowledge for teaching – What makes it special? *Journal of Teacher Education, 59*(5), 389–407.

Chai, C. S., Koh, J. H. L., & Tsai, C. C. (2010). Facilitating preservice teachers' development of technological, pedagogical, and content knowledge (TPACK). *Journal of Educational Technology & Society, 13*(4), 63–73.

ChanLin, L. J. (2008). Technology integration applied to project-based learning in science. *Innovations in Education and Teaching International, 45*(1), 55–65.

Creswell, J. W. (2008). *Educational research: Planning, conducting, and evaluating quantitative and qualitative research* (3rd ed.). Upper Saddle River, NJ: Pearson.

Creswell, J. W., & Plano Clark, V. L. (2011). *Designing and conducting mixed methods research* (2nd ed.). Los Angeles: Sage.

Ebner, M., Lienhardt, C., Rohs, M., & Meyer, I. (2010). Microblogs in higher education – A chance to facilitate informal and process-oriented learning? *Computers & Education, 55*(1), 92–100.

Edelson, D. C. (2001). Learning-for-use: A framework for the design of technology-supported inquiry activities. *Journal of Research in Science Teaching, 38*(3), 355–385.

Flick, U. (2002). *An introduction to qualitative research*. London: Sage.

Graham, C. R. (2011). Theoretical considerations for understanding technological pedagogical content knowledge (TPACK). *Computers & Education, 57*(3), 1953–1960.

Guba, E., & Lincoln, Y. S. (1989). *Fourth generation evaluation*. Newbury Park, CA: Sage.

Hofer, B. K. (2000). Dimensionality and disciplinary differences in personal epistemology. *Contemporary Educational Psychology, 25*(4), 378–405.

Hsu, Y. S., Wu, H. K., & Hwang, F. K. (2007). Factors influencing junior high school teachers' computer-based instructional practices regarding their instructional evolution stages. *Educational Technology & Society, 10*(4), 118–130.

Hughes, J. (2005). The role of teacher knowledge and learning experiences in forming technology-integrated pedagogy. *Journal of Technology and Teacher Education, 13*(2), 277–302.

Jang, S. J., & Tsai, M. F. (2012). Exploring the TPACK of Taiwanese elementary mathematics and science teachers with respect to use of interactive whiteboards. *Computers & Education, 59*(2), 327–338.

Jimoyiannis, A. (2010). Designing and implementing an integrated technological pedagogical science knowledge framework for science teachers' professional development. *Computers & Education, 55*(3), 1259–1269.

Kay, R. H., & LeSage, A. (2009). Examining the benefits and challenges of using audience response systems: A review of the literature. *Computers & Education, 53*(3), 819–827.

Keating, T., & Evans, E. (2001, March). *Three computers in the back of the classroom: Pre-service teachers' conceptions of technology integration*. Paper presented at the annual meeting of the American Educational Research, Seattle, WA.

Kim, M. C., & Hannafin, M. J. (2011). Scaffolding problem solving in technology-enhanced learning environments (TELEs): Bridging research and theory with practice. *Computers & Education, 56*(2), 403–417.

Klassen, R. M., Tze, V. M. C., Betts, S. M., & Gordon, K. A. (2010). Teacher efficacy research 1998–2009: Signs of progress or unfulfilled promise? *Educational Psychology Review, 23*(1), 21–43.

Koehler, M. J., & Mishra, P. (2005). What happens when teachers design educational technology? The development of technological pedagogical content knowledge. *Journal of Educational Computing Research, 32*(2), 131–152.

Koehler, M. J., Mishra, P., & Yahya, K. (2007). Tracing the development of teacher knowledge in a design seminar: Integrating content, pedagogy and technology. *Computers & Education, 49*(3), 740–762.

Krajcik, J., McNeill, K. L., & Reiser, B. J. (2008). Learning-goals-driven design model: Developing curriculum materials that align with national standards and incorporate project-based pedagogy. *Science Education, 92*(1), 1–32.

Kramarski, B., & Michalsky, T. (2010). Preparing preservice teachers for self-regulated learning in the context of technological pedagogical content knowledge. *Learning and Instruction, 20*(5), 434–447.

Lee, M. H., & Tsai, C. C. (2010). Exploring teachers' perceived self-efficacy and technological pedagogical content knowledge with respect to educational use of the World Wide Web. *Instructional Science, 38*(1), 1–21.

Lee, S. W. Y., Tsai, C. C., Wu, Y. T., Tsai, M. J., Liu, T. C., Hwang, F. K., et al. (2011). Internet-based science learning: A review of journal publications. *International Journal of Science Education, 33*(14–15), 1893–1925.

Lin, T. C., Tsai, C. C., Chai, C. S., & Lee, M. H. (2013). Identifying science teachers' perceptions of technological pedagogical and content knowledge (TPACK). *Journal of Science Education and Technology, 22*(3), 325–336.

Linn, M. C. (2003). Technology and science education: Starting points, research programs, and trends. *International Journal of Science Education, 25*(6), 727–758.

Linn, M. C., Clark, D., & Slotta, J. D. (2003). WISE design for knowledge integration. *Science Education, 87*(4), 517–538.

Lorr, M. (1983). *Cluster analysis for social scientists: Techniques for analyzing and simplifying complex blocks of data*. San Francisco: Jossey-Bass.

Lumpe, A., Czerniak, C., Haney, J., & Svetlana, B. (2012). Beliefs about teaching science: The relationship between elementary teachers' participation in professional development and student achievement. *International Journal of Science Education, 34*(2), 153–166.

Lundeberg, M. A., Bergland, M., Klyczek, K., & Hoffman, D. (2003). Using action research to develop preservice teachers' beliefs, knowledge and confidence about technology. *Journal of Interactive Online Learning, 1*(4), 1–16.

Mäkitalo-Siegl, K., Kohnle, C., & Fischer, F. (2011). Computer-supported collaborative inquiry learning and classroom scripts: Effects on help-seeking processes and learning outcomes. *Learning and Instruction, 21*(2), 257–266.

Mishra, P., & Koehler, M. J. (2006). Technological pedagogical content knowledge: A framework for teacher knowledge. *Teachers College Record, 108*(6), 1017–1054.

Mishra, P., Koehler, M. J., & Kereluik, K. (2009). Looking back to the future of educational technology. *TechTrends, 53*(5), 48–53.

Niederhauser, D. S., & Stoddart, T. (2001). Teachers' instructional perspectives and use of educational software. *Teaching and Teacher Education, 17*(1), 15–31.

Niess, M. L. (2005). Preparing teachers to teach science and mathematics with technology: Developing a technology pedagogical content knowledge. *Teaching and Teacher Education, 21*(5), 509–523.

Palmer, D. H. (2002). Factors contributing to attitude exchange amongst preservice elementary teachers. *Science Education, 86*(1), 122–138.

Parkinson, J. (1998). The difficulties in developing information technology competencies with student science teachers. *Research in Science & Technological Education, 16*(1), 67–78.

Sandholtz, J. H., Ringstaff, C., & Dwyer, D. C. (1997). *Teaching with technology: Creating student-centered classrooms*. New York: Teachers College Press.

Schmidt, D. A., Baran, E., Thompson, A. D., Mishra, P., Koehler, M. J., & Shin, T. S. (2009). Technological pedagogical content knowledge (TPACK): The development and validation of an assessment instrument for preservice teachers. *Journal of Research on Technology in Education, 42*(2), 123–149.

Serin, O. (2011). The effects of the computer-based instruction on the achievement and problem solving skills of the science and technology students. *Turkish Online Journal of Educational Technology, 10*(1), 183–201.

Shulman, L. S. (1986). Those who understand: Knowledge growth in teaching. *Educational Researcher, 15*(2), 4–14.

Shulman, L. S. (1987). Knowledge and teaching: Foundations of the new reform. *Harvard Educational Review, 57*(1), 1–22.

Strauss, A., & Corbin, J. (1990). *Basics of qualitative research: Grounded theory procedures and techniques*. Newbury Park, CA: Sage.

Suthers, D. D. (2006). Technology affordances for intersubjective meaning making: A research agenda for CSCL. *Computer-Supported Collaborative Learning, 1*(3), 315–337.

Thompson, A. D., & Mishra, P. (2008). Breaking news: TPCK becomes TPACK! *Journal of Computing in Teacher Education, 24*(2), 38–64.

van Driel, J. H., Beijaard, D., & Verloop, N. (2001). Professional development and reform in science education: The role of teachers' practical knowledge. *Journal of Research in Science Teaching, 38*(2), 137–158.

Voogt, J., Fisser, P., Pareja Roblin, N., Tondeur, J., & van Braak, J. (2013). Technological pedagogical content knowledge – A review of the literature. *Journal of Computer Assisted Learning, 29*(2), 109–121.

Web, M., & Cox, M. (2004). A review of pedagogy related to information and communications technology. *Technology, Pedagogy & Education, 13*(3), 235–286.

Yeh, Y. F., Hsu, Y. S., Wu, H. K., Hwang, F. K., & Lin, T. C. (2014). Developing and validating technological pedagogical content knowledge – Practical (TPACK-Practical) through the Delphi survey technique. *British Journal of Educational Technology, 45*(4), 707–722. doi:10.1111/bjet.12078.

Yung, B. H. W., Zhu, Y., Wong, S. L., Cheng, M. W., & Lo, F. Y. (2013). Teachers' and students' conceptions of good science teaching. *International Journal of Science Education, 35*(14), 2435–2461.

Yurdakul, I. K., Odabasi, H. F., Kilicer, K., Coklar, A. N., Birinci, G., & Kurt, A. A. (2011). The development, validity and reliability of TPACK-deep: A technological pedagogical content knowledge scale. *Computers & Education, 58*(3), 964–977.

Part II
The Transformative Model of TPACK

Chapter 4
Rubrics of TPACK-P for Teaching Science with ICTs

Yi-Fen Yeh, Sung-Pei Chien, Hsin-Kai Wu, and Ying-Shao Hsu

Advances in information communication technologies (ICTs) have diversified teacher instruction. The appropriateness of representation selections and learning activity designs involving ICTs is determined by teachers' technological pedagogical content knowledge-practical (TPACK-P), a knowledge construct transformed and reinforced through different tasks in teaching. This study developed rubrics for evaluating preservice teachers' TPACK-P, according to the proficiency levels and features identified by in-service teachers. We collected lesson plans and microteaching video clips of seven preservice teachers in order to verify the rubrics and explore how their TPACK-P was demonstrated in lesson plans and microteaching. Results revealed that the preservice teachers' performances on lesson planning and microteaching were similar, with discrepancies of +/− 1 level on the rubrics. Their performances on teaching with ICTs were comparatively better in curriculum design and enactment than on assessment. It may not be difficult for preservice teachers to implement ICTs, but the real challenges are to use ICTs with considerations of students, content, and pedagogy. Teacher education programs are advised to pay attention to how meaningfully ICTs are used to support instruction, rather than simply counting the number of times ICTs are used.

Y.-F. Yeh (✉)
Science Education Center, National Taiwan Normal University, Taipei, Taiwan
e-mail: yyf521@ntnu.edu.tw

S.-P. Chien • H.-K. Wu • Y.-S. Hsu
Graduate Institute of Science Education, National Taiwan Normal University, Taipei, Taiwan
e-mail: cellist_2@hotmail.com; yshsu@ntnu.edu.tw

© Springer Science+Business Media Singapore 2015
Y.-S. Hsu (ed.), *Development of Science Teachers' TPACK*,
DOI 10.1007/978-981-287-441-2_4

4.1   Introduction

Teaching is a profession in which teachers transform content knowledge into learnable events and help students establish their own knowledge (Shulman, 1987). Ideally, beginning teachers are expected to have not only the necessary knowledge for instruction but also the competence to solve problems with comprehensive considerations and appropriate solutions. Teacher education programs that contextualize preservice teachers in classroom practices or oversee their student-teaching experiences have been found to develop better teachers and lead to higher teaching placement rates and student learning outcomes (Boyd, Grossman, Lankford, Loeb, & Wyckoff, 2009). Therefore, high-quality teaching demonstrations where information communication technologies (ICTs) are infused will enhance teachers' knowledge about teaching with ICTs, and teaching experiences will help them integrate these proficiencies and knowledge.

Teachers' professional knowledge is a type of craft knowledge (Shulman, 1986). Teaching experiences can be viewed as a catalyst to transform teachers' academic knowledge into practical knowledge with which teachers facilitate their instruction. In-service teachers refine their professional knowledge through actual classroom teaching experiences, while preservice teachers can only develop such professional knowledge from microteaching, student teaching, clinical experiences, or teaching internships. Since reflection-on-action can promote reflection-in-action, we believe that developing performance rubrics to evaluate preservice teachers' knowledge and actions in teaching with ICTs will be helpful to teacher education. Analyzing preservice teachers' lesson plans and corresponding microteaching video clips will help us to further elaborate and verify the rubrics.

4.1.1   Gaps Between Preservice Teachers' and In-Service Teachers' Professional Knowledge

The professional knowledge necessary for technology-infused instruction has been identified as technological pedagogical content knowledge (TPACK; Mishra & Koehler, 2006). TPACK refers to an integrative knowledge construct that is composed of teachers' knowledge about content, pedagogy, and technology as well as the intersecting components of these major knowledge categories (Mishra & Koehler). Cox and Graham (2009) emphasized the dynamic and transactional features of TPACK by defining it as "teachers' knowledge of how to coordinate subject- or topic-specific activities with topic-specific representations using emerging technologies to facilitate student learning" (p. 64). Hence, teachers' TPACK proficiency does not rest on knowledge alone but also on how responsive and flexible teachers are when making and enacting decisions about technology use in instructional situations.

Novice teachers are expert learners but not necessarily expert teachers-to-be (Shulman, 1987). Instructional experiences are the main attributors to the gaps between learning and teaching, as well as between novice teachers and experienced teachers (van Driel, Verloop, & de Vos, 1998). Experiences provide the interactive and constructive contexts necessary to transform existing ideas, develop new insights and applications, and steadily shape a much more well-defined and unified knowledge construct of pedagogical content knowledge (PCK) and TPACK (Angeli & Valanides, 2009; Gess-Newsome, 1999; Gess-Newsome & Lederman, 1993; Magnusson, Krajcik, & Borko, 1999). For example, science teachers can transform their PCK through repeated engagement of teacher-directed explanations or student-directed knowledge construction. ICTs can be tools to further accommodate both types of instruction if teachers consider topics, technological affordances, learners, and pedagogies all together (Angeli & Valanides, 2009). Designing lessons, microteaching, and teaching internships can be good rehearsals for preservice teachers to practice and transition the knowledge they acquire from individual courses into professional knowledge that is personally consolidated and dynamically situated within instructional contexts. Artifacts from these experiences can then be used to document the development of their professional knowledge.

Cox and Graham (2009) analyzed a case study regarding how a scientist who was an adaptive technology user taught her undergraduate and graduate geology classes. As a scientist, she used ice core drills, ice-penetrating radar, computers, and 3-D models to sample data, explore the structure of a glacier, analyze data, and reconstruct 3-D models of glaciers. As a teacher, she used PowerPoint flexibly to help her stay organized and focused when presenting information and delivering graphic representations. She juxtaposed pictures of U- and V-shaped valleys to help students view types of erosion and trigger discussions, but she found that this type of lecture did not facilitate her students' understanding of glacial advances and retreats. Therefore, she chose to use in her classroom a model she had built for research purposes and allowed students to discover concepts and run simulations by manipulating variables (e.g., temperature, precipitation). She also used whiteboards when teaching equations that required variable manipulation and additional input. This exemplary case demonstrated excellent technological content knowledge (TCK) and technological pedagogical knowledge (TPK; Cox & Graham, 2009) within which tool affordances are tacitly used to support subject matter, student needs, and pedagogical strategies employed for better learning results (Angeli & Valanides, 2009). The thoughtful design thinking that this exemplary teacher engaged in her instruction demonstrates the TPACK that digital-age teachers should be expected to develop. TPACK should be also viewed as an *in progress* body of knowledge, considering the fact that new technologies keep coming out and these technologies can be implemented in versatile teaching and learning activities.

Under the definition that TPACK is experience based, inexperienced preservice teachers are less proficient than experienced in-service teachers due to a lack of experiences designing and enacting technology-infused instruction. However, many of these novice teachers are likely to be receptive to ICT-based instructional applications since they are digital natives and members of the *Net Generation*. A series of teaching practica that prepares preservice teachers for dealing with different instructional tasks and classroom environments can be useful for filling in knowledge–practice gaps.

4.1.2 Rubrics for Evaluating Teachers' PCK and TPACK

The feature of *situativeness* makes teachers' PCK and TPACK not only difficult to develop but also complicated to evaluate outside of an actual learning–teaching context. Thus, performance-based TPACK measures offer more valid and reliable measurements compared to the often-used, self-report approach. For example, Abbitt (2011) evaluated preservice teachers' solutions to teaching scenarios and design-based activities in order to document preservice teachers' TPACK.

Microteaching is an activity that has long been used in teacher education programs since it offers "both beginning and advanced teachers excellent opportunities to plan and practice a wide array of new instructional strategies" (Orlich et al., 1990, p. 169). Some preservice teachers may view such instructional rehearsals as *fake teaching* that offers only structured/controlled teaching experiences, but others have a positive attitude toward such practices of lesson planning and enactment (Bell, 2007; Metcalf, 1993; Pauline, 1993). In fact, "interactive teaching [experiences] in settings of reduced complexity" (Grossman & McDonald, 2008, p. 190) help preservice teachers gain new insights in content, pedagogy, and technology that are different from those gained from books or lectures. Explicit learning objectives, feedback, and reflection on microteaching experiences are necessary if teacher educators want preservice and practicing teachers to gain experience-based professional knowledge from teacher education or professional development programs (Hatfield, 1989). Rubrics or criteria that list goals and expected instructional behaviors may function not only as follow-up evaluative tools for teaching performance but also as benchmarks to pursue before and during microteaching.

The focus of teacher evaluations varies with how teacher educators view science instruction and the related uses of ICTs. Earlier rating scales for science teachers' microteaching performances (without ICT implementation as a requirement) focused mainly on the quality of their traditional lecture and inquiry guidance in the lesson's opening, body, and ending (Pauline, 1993); the ICTs were viewed simply as information sources and communication devices. More recently, as ICTs have permeated more thoroughly into educational contexts, rubrics for ICT-infused instruction have paid greater attention to whether instructional objectives or ICTs were considered or implemented (Mitchem, Wells, & Wells, 2003). Later, the evaluation focus switched to how ICTs assisted the teacher's instruction (Niess, 2005) or how ICTs influenced the design of the lesson (Angeli & Valanides, 2005). After the introduction of the concept of TPACK in teacher education, the criteria for evaluating teachers' instructional artifacts switched to validating those that focused on mapping compatibility among curriculum goals, technology selections, and instructional strategies (Harris, Grandgenett, & Hofer, 2010) and whether the ICTs supported learners as active agents in a cooperative knowledge construction process (Jonassen, Howland, Marra, & Crismond, 2008; Lee, Chai, & Koh, 2012). Therefore, the evaluation of teachers' ICT-infused instruction should emphasize not only whether and what ICTs are used but also how they are used.

Considering that teaching is a knowledge- and performance-based behavior within an instructional context, the progression or development of teachers' understanding of how to teach with ICT usually begins with knowing and moves to different maturity levels in implementation (Moersch, 1995; Russell, 1995). Proficiency in such an awareness–implementation progression moves through levels of entry, adoption, adaptation, appropriation, and invention (Sandholtz, Ringstaff, & Dwyer, 1997) or stages of recognizing, accepting, adopting, exploring, and advancing (Niess, 2012; Niess et al., 2009). Chapter 3 of this book presents the features of in-service science teachers' developmental progression in using technology to assist their instruction. However, are the features identified in teachers' performances in real classrooms also descriptive of preservice teachers' performances with regard to lesson planning and microteaching?

4.2   Teacher Education and Professional Learning

Measuring how well teachers know and apply TPACK-P can be difficult. Teachers' proficiency in ICT implementation may vary not only across instructional activities but also across the behaviors to be evaluated. For example, teachers may be knowledgeable in ICT-infused representations but not in ICT-infused assessments. Teachers may plan well for teaching with ICTs, but they may not enact these plans properly or perform equally well in their instruction. Unpredictable responses from students can be challenges especially for novice teachers. Therefore, the focus of this chapter is to develop rubrics to evaluate preservice teachers' TPACK and report their distinctive features in curriculum designing and teaching from their microteaching experiences.

4.2.1   Rubrics for Lesson Plans and Microteaching

Science educators and expert teachers have identified TPACK-P as a knowledge construct with eight dimensions that teachers encounter or deal with throughout their teaching, covering the knowledge domains of assessment, planning and designing, and teaching practices (see Chap. 2). Interviews with in-service teachers who had experience teaching with technology were used to identify the levels of teacher proficiency in TPACK-P and the typical features of teachers at these levels (see Chap. 3). These findings provided a foundation for generating rubrics to be used with preservice science teachers.

Due to gaps in the knowing–applying spectrum, the rubrics for performance-based evaluations require modifications to the TPACK-P framework, which was originally developed from teachers' perceptions of technology uses in instruction (Chaps. 2 and 3). First, we modified the dimension *planning ICT-infused curriculum* to be *B. constructing ICT-infused curricula* so that teachers would be evaluated by how well they planned and physically implemented ICTs with proper learning, goal setting, and completeness. Previous dimensions of *using ICTs to understand subject content* and *applying ICTs to instructional management* are not involved in actual lesson design or enactment behaviors. Second, the knowledge dimensions of the original framework were constructed based on the idea of teaching procedures, which in nature lacked overall evaluations regarding how well these instructional uses were meaningfully connected and carried over. Therefore, we added the dimension *E. providing explicit guidance when using ICTs* to the knowledge domain "Teaching Practice" to evaluate how explicitly teachers orally guided and engaged their students in ICT-infused learning. After identifying and making the necessary modifications, we deemed seven knowledge dimensions to be critical to lesson planning and microteaching in which teachers demonstrate their TPACK-P. They are:

- A. Using ICTs to understand students and assist with their learning
- B. Constructing ICT-infused curricula
- C. Using ICT-supported representations to present instructional representations
- D. Employing ICT-integrated teaching strategies
- E. Providing explicit guidance when using ICTs
- F. Preparing or troubleshooting technical problems in teaching contexts
- G. Using ICTs to assess students

Teachers' proficiency levels were also modified since teachers' perceptions (Chap. 3) could not be fully reflective of their performance. Therefore, by leaving out the perceptional level that involves no actual implementation (Level 0 – no idea), we used four performance-based levels to evaluate preservice teachers' microteaching performances (i.e., lack of use, simple adoption, infusive application, student-centered applications). Performance features identified from teachers with different proficiency levels in lesson planning and microteaching were also included as representative features and descriptors for each rubric item.

4.2.2 Data Collection and Analysis

The preservice teachers who participated in this study were students enrolled in the Physics Teaching Practicum. The first part of the course developed these preservice physics teachers' PCK, and the second part developed their TPACK; each part required them to submit lesson plans and do microteaching. The task of teaching with ICTs required these preservice teachers to teach with physics learning applications (APPs) on their tablets. These APPs were developed by groups of in-service teachers, the course professor, and researchers (Chap. 5). The teacher educator in this course prepared these physics teachers-to-be through rationalizing why and how the physics APPs were designed and followed up with constructive comments regarding these preservice teachers' instructional designs and microteaching practices. We collected seven sets of lesson plans and their corresponding microteaching videos from volunteers for use in verifying the rubrics.

Two of the authors randomly selected two preservice teachers as exemplar cases. First, the two coders carefully scrutinized these two sets of lesson plans and microteaching videos according to the features of the four proficiency levels that were identified from in-service teachers' self-reports in interview data (Chap. 3). They discussed what items might need to be deleted or added to make the rubrics more discriminating. Another round of separate coding for the same two sets of instructional artifacts was conducted and followed with a discussion of the discrepancies and rubric modifications. Second, the two coders used the finalized version of the rubrics to rate all seven sets of lesson plans and microteaching videotapes separately. Table 4.1 lists (a) the rubric items for different aspects of the teachers' instruction that deserve to be effectively evaluated and (b) the features of the preservice teachers' performances on lesson plans and microteaching videos. Since proficiency levels are on a continuum (i.e., Level 1 to Level 4), teachers identified at a higher proficiency level were assumed to have acquired certain features at equivalent or lower levels. The interrater reliability was 82 % in the first round of separate ratings; discussion of discrepancies in the ratings resulted in consensus agreement.

4.3   Preservice Teachers' Lesson Plans and Microteaching

Preservice teachers' performances on lesson plans and microteaching were rated from lack of use (1) to student-centered applications (3). The seven teachers' performances on the seven dimensions of ICT implementation in educational contexts (planning and microteaching) were then summarized.

4.3.1   Patterns of Preservice Teachers' ICT Use

Figure 4.1 shows the seven preservice teachers' performance proficiencies across the various dimensions of TPACK-P. By plotting each case teacher's proficiency profile through web graphics, within- and between-case comparisons became easily made. We also report their proficiency distributions according to dimensions, levels, and tasks in Table 4.2. Summary assertions from cross-case comparisons were made later to reveal major findings from these preservice physics teachers' performances.

Some teachers may have demonstrated good TPACK in their lesson plans but did not necessarily demonstrate similar proficiencies in their microteaching (Case 5). Cases 1 and 4 demonstrated reasonable performances in terms of lesson planning

Table 4.1 Rubrics for evaluating preservice teachers' performances in ICT-infused instruction

	1 – Lack of use	2 – Simple adoption	3 – Infusive application	4 – Student-centered application
Knowledge of assessments				
A. Using ICTs to understand students and assist with their learning	Unable to use any ICTs to probe students' knowledge	Uses ICTs as probing tools (i.e., information presentation) when trying to understand students' knowledge and/or facilitate their learning	Observes students' learning characteristics from their interactions with ICTs. Teacher's instruction is designed and enacted based on students' learning characteristics	Uses the interactive response system (IRS) to understand students and engage student-centered, ICT-infused instruction to let students self-construct their learning
Knowledge of planning and designing				
B. Constructing ICT-infused curricula	Sets goals for ICT implementation as motivation booster (no evidence)	Arranges ICTs to support teacher's instruction, helping students to achieve benchmark learning goals	Arranges different ICT uses to consolidate students' concept understanding and clarify their misconceptions	Designs ICT-infused curricula as tasks that encourage students to do scientific explorations or concept construction
C. Using ICT-supported representations to present instructional representations	Uses traditional representations (e.g., drawing) to present target concepts (no evidence)	Uses ICTs to present scientific concepts that are abstract, dynamic, and micro- or macroscopic for students to visualize	Engages students to interact with ICT-infused representations or instruments when conceptualizing science concepts or scientific thinking	Designs/engages learning tasks that are infused with ICT representations or instruments for students to do virtual experiments or scientific explorations
D. Employing ICT-integrated teaching strategies	Uses chalk-and-talk lecture when implementing ICTs in class	Uses pedagogical strategies (e.g., questioning) that increase students' motivations and understanding of the content when using ICTs in instruction	Uses pedagogical strategies that facilitate students' comprehension or discussion of the target concepts when learning with ICTs (e.g., P-O-E)	Uses pedagogical strategies to help students personally or collaboratively accomplish ICT-infused learning tasks (e.g., inquiry)

Knowledge of teaching practices

E. Providing explicit guidance when using ICTs	*Plan*: lists instruction with short steps *Teaching*: offers unclear instruction guiding students to understand or manipulate ICTs	*Plan*: briefly lists objectives, content, and procedures *Teaching*: uses understandable language to explain the target scientific concepts with ICT implementations, but greater coherence or logic needed	*Plan*: lists the guiding questions or potential concepts to be included *Teaching*: uses comprehensive, learning-facilitated language to conceptualize the target scientific concepts with ICT implementations	*Plan*: lists possible questions or reactions students may have *Teaching*: is responsive to students' learning needs with explicit, thought-provoking guidance, when helping students construct the target scientific concepts with ICT implementations
F. Preparing or troubleshooting technical problems in teaching contexts	Unable to continue the course (no evidence)	Has no backup plans or is unable to solve technical problems (i.e., implies she/he may go back to chalk-and-talk lectures)	Engages solutions or alternatives that are technologically based (i.e., posts e-files for students to download after school)	Demonstrates her/his competence in troubleshooting without interrupting courses (e.g., prepares hard copies for Internet troubles)

Knowledge of assessments

G. Using ICTs to assess students	Unable to use any ICTs to assess students' learning	Evaluates students' learning with formative or summative assessments that are implemented with ICTs (e.g., multimedia)	Evaluates students with use of electronic assessment systems that offer responsive feedback or keep learning records (e.g., simulation, IRS, electronic platform)	Uses ICT-infused assessments that engage students to demonstrate their scientific thinking or ability to construct models from the built-in simulations or dataset

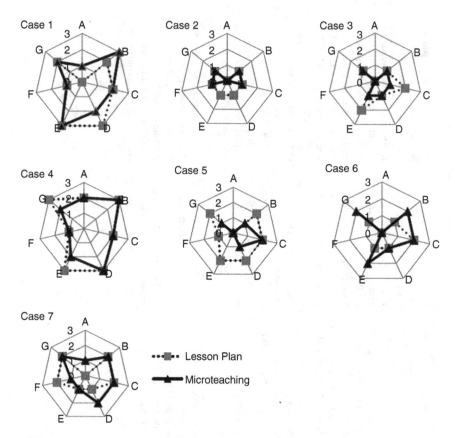

Fig. 4.1 Preservice teachers' performances in the seven dimensions of TPACK-P

Table 4.2 Distribution of the seven preservice teachers by their proficiency level

	Dimension of TPACK-P													
	A		B		C		D		E		F		G	
Level	LP	MT	LP	MT	LP	MT	LP	MT	LP	MT	LP	MT	LP	MT
0	**6**	**4**	0	0	0	0	0	1	0	**2**	2	**3**	0	0
1	0	2	**3**	**3**	1	2	**4**	**3**	2	2	**5**	**4**	**3**	**3**
2	1	1	**3**	2	**6**	**5**	1	2	**3**	2	0	0	**3**	**4**
3	0	0	1	2	0	0	2	1	2	1	0	0	1	0

Note. Numbers in **bold** are the proficiency levels that most teachers achieved
LP lesson plan, *MT* microteaching

(LP) and microteaching (MT), but both had room for further development. Case 2 demonstrated a reasonably consistent poor performance across all dimensions in both tasks. The other cases (3, 5, 6, and 7) demonstrated moderate but varying performances. Generally speaking, they had better or equally proficient performances in LP than MT for all cases and dimensions. Almost all of the discrepancies were within one proficiency level in LP and MT performance, except for Dimension E (engaging explicit guidance when using ICTs) in Case 5. Accordingly, preservice teachers' proficiencies between planning (i.e., lesson plans) and teaching (i.e., microteaching) were quite consistent.

Almost all of the seven preservice teachers were good at Dimension C (using ICT-based representations to present instructional representations); they were rated at Level 2 in both LP and MT. Three of the teachers used ICTs to understand and assist with their learning (Dimension A), and all used ICTs to assess student learning after teaching certain concepts (Dimension G). In fact, these three teachers had higher scores in the other dimensions as well. Therefore, it was inferred that the preservice teachers who were able to implement ICTs to understand student learning intended to further elaborate their instruction with different tools.

4.3.2 Teaching Behaviors with ICTs

These seven preservice teachers displayed a variety of behaviors in planning and implementing ICTs in their physics instruction. They used ICTs and engaged pedagogical strategies to accomplish ICT implementation. Figure 4.2 maps the tasks in the plans as well as ICT selections and pedagogical strategies these preservice teachers engaged. Links (lines) connect the task, tool, and strategy nodes (boxes) where evidence was identified from their performances.

- Images were mainly used to present scientific phenomena; video clips (i.e., news clips) were used to increase students' learning motivation and curiosity.
- Pedagogical strategies made ICT implementation more educationally meaningful. These preservice teachers usually used images along with predict-observe-explain (POE) guidance when they wanted their students to observe and scrutinize

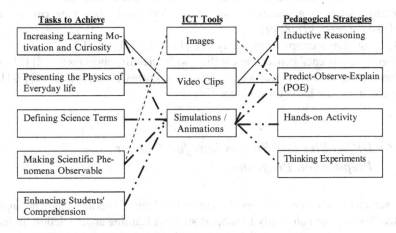

Fig. 4.2 Mapping of learning tasks, ICT tools, and pedagogical strategies

the presented static phenomena. Besides POE, inductive reasoning allowed them to guide students to explain the dynamic phenomena or experimental results that the video clips presented.

- They used simulations in their instruction and engaged students to use them as well. For example, simulation-based APPs were used as initial activities to trigger students' curiosity and as tools to enhance students' inquiry ability through variable manipulation and hypothesis testing.

Overall, the ways these preservice teachers used ICTs to construct their students' knowledge of science showed their TPACK or their TPACK-practical (TPACK-P). Flexible uses of pedagogical strategies were found when these preservice teachers used simulations to teach physics, which led to the fulfillment of different instructional goals. However, comparatively limited arrays of pedagogical strategies and tasks were engaged when these preservice teachers used images and video clips to assist their instruction.

In order to better understand how these preservice teachers used ICTs to assist their instruction, we detailed Case 4's lesson plan (see Appendix). This lesson plan describes how he planned and delivered instruction regarding the concept of friction with the use of ICTs and the physics APP. His microteaching behaviors were quite consistent with what he planned to do in his lesson plan (see web graphic in Fig. 4.1), except for two (E & G) of the seven dimensions where his LP was higher than his MT. As shown in Fig. 4.1, his teaching performance in all dimensions except using ICTs to assess his students (Dimension F) categorized him as a teacher with TPACK-P proficiency ranked Level 2 or above. His lesson plan could be used as an exemplar showing how simulations and other ICTs can be woven smoothly throughout warm-up, content delivery, and student practice. His 30-min class included a total of 11 instructional episodes (in square brackets, as numbered in the Appendix). A summary of the main features identified from his flexible simulation usages for specific purposes and episodes include:

- Gaining experimental data from repeated hypotheses tested in simulations. [7] [8] [0]
- Offering simulation results for students to compare when they constructed related concepts. [7] [8] [9]
- Demonstrating examples from daily life. [1] [2] [3] [4] [5] [6]
- Triggering students' conceptual conflicts with scientific phenomena. [2] [3] [6]
- Concretizing abstract concepts or thinking experiments. [1] [3] [6] / [7] [8] [9] [10]
- Organizing main ideas for students. [11]

4.3.3 Difficulties and Suggestions for Teacher Preparation Programs

The exemplar lesson plan provides ideas regarding how simulations might be implemented flexibly and coherently to support students learning about friction. In fact, some of the preservice teachers implemented the simulation-based APPs merely as a course requirement and lacked a full consideration of all factors involved in teaching and learning that successful ICT-infused instruction requires. Though some participants may point out the errors and misguiding concepts embedded in the animations and simulations they used (e.g., Case 7), some still need to anticipate what alternative concepts their students may have after interacting the simulation (i.e., Case 4) and then make necessary clarifications. Offering concise instruction to guide students to make observations or manipulate simulations is important, but it needs time to fully work (e.g., Cases 2 and 5). Some of the participating preservice teachers showed their unpreparedness with the ICT implementation, which later led to instruction interruption. For example, Case 1 searched for the "replay" button in the APP exhibiting projectile motion, but there was no such button (although there was in other APPs). Case 6 did not plan ahead as evidenced by the tablet PCs that he offered to his students did not have the latest Flash® upgrades. Other problems included bad connections between tablets and the projector that teachers should easily sort out prior to teaching or their underestimation of the time students might need when manipulating the simulations.

4.4 Final Thoughts and Implementations

Pineau (1994) noted that "[h]istorically there are strong links between performance methodology and methods of teacher training. 'Rehearsing' one's teaching personae is well established in educational literature and practice" (p. 17). The rubrics we used to evaluate teachers' instructional artifacts (lesson plans and videotapes of microteaching sessions) reflected how we conceptualized TPACK-P and what we expected teachers to achieve when teaching with ICTs. We believe that the more these infused ICT uses are appropriately planned, applied, and reflected upon from different aspects of teaching, the more proficient teachers will be in developing their TPACK-P (Yeh, Hsu, Wu, Hwang, & Lin, 2014). Engaging preservice teachers in tasks of instructional design and enactment will add a practice-oriented dimension to their TPACK, transforming it into TPACK-P.

Physics is a subject area that demands more – compared with other science subjects – of learners' cognitive development and abstract thinking ability. Physics teachers tend to spend time constructing students' modeling abilities when teaching inquiry classes (Breslyn & McGinnis, 2011) and helping students comprehend abstract concepts through formula verifications (Mulhall & Gunstone, 2008). Expert teachers in physics rated ICT applications as less important, especially when compared to science teachers with backgrounds in biology, chemistry, and earth science (Yeh et al., 2014). However, simulations and microcomputer-based laboratories are helpful in teaching abstract physics concepts (Perkins et al., 2006). These tools have been found to enhance students' and preservice teachers' learning of physics if they are intentionally designed and used to enhance learners' cognition and conceptual understanding (Bernhard, 2003). Furthermore, no matter how well designed or facilitative these simulation-based APPs may be to students' science learning, we should not forget that teachers' guidance also determines how and what students learn from these instructional technologies.

Pedagogical knowledge is fundamentally important to the professional development of teachers; it is usually the knowledge that preservice teachers lack and likely are unaware that they lack. *Lesson planning* includes decision making regarding appropriate activities, sequencing the lessons and learning opportunities within a unit of study, and selecting technological tools for students; *lesson implementation* includes the enactment of instructional preparation, appropriate laboratory demonstrations or experiments, and scientific modeling for students. Good multimedia uses require pedagogical rationality (Mouza, 2003). TPK was found to make the most critical contribution to the success of preservice teachers' instruction (Jaipal & Figg, 2010). In Taiwan, the teacher recruitment system overemphasizes how well teacher candidates organize curricula and explain concepts understandably to students in chalk-and-talk format when they do teaching demonstrations. After these teachers are hired, they are expected and encouraged to teach with ICTs when schools have a goal of promoting technology-infused instruction. These conflicting teacher expectancies, recruitment procedures, and future school plans may either leave out teachers who are able to offer explicit instruction with ICTs or recruit teachers who are satisfied with their chalk-and-talk instruction but unwilling or not confident enough to embrace science instruction with ICTs. It is also possible that even teachers who are frequent users of technology in daily life or in lecturing still encounter difficulties offering explicit instructions geared to help students manipulate ICTs.

Findings and patterns observed regarding how these teachers performed when using simulation-based APPs in their instruction could be useful to the design of future teacher education; however, we believe more data would further validate or expand the rubrics and our understanding of TPACK-P. Rationalizing the uses of ICT tools can be a useful strategy in teacher preparation. Therefore, we suggest that teacher preparation programs not be limited to helping teachers deliver subject content with ICTs; furthermore, how ICT applications assist learning evaluation and assist learning progress should be implemented into programs of teacher education and professional development. Findings indicated in-service teachers' preference for paper-and-pencil assessments even when they did teach with ICTs (Chap. 3) and preservice teachers' uses of ICT-infused assessments to evaluate students' learning but ignorance of students' individual differences and prior knowledge (this chapter). Using ICTs to assist science content delivery and students' inquiry ability construction is important, but knowing students and keeping track of their learning progress throughout their science learning would make teaching and learning effective and efficient.

Appendix: Lesson Plan Exemplar (Case 4)

Theme: classical mechanics	Topic: static friction and dynamic friction
Target students: Grade 9	Duration: 45 min
Curriculum: handouts for junior high school students	Tools: tablet PCs, projectors

(continued)

Learning objectives:

1. Get students to be interested in the content (i.e., friction)

2. Let students find answers from the experiments

Content analysis:

1. Motivate students to learn physics by displaying videos and referring to similar cases in daily life situations

2. Teach students by making connections between subject content and experiment results

Presentation of the concept of friction (15 min)

[1] Display a commercial video about tires. Familiarize students with the course topic by asking them what the commercial tries to convey. The commercial shows that a man suddenly depresses the brakes in his car once he sees a car rolling into the road. Later, he gets angrily out of the car and tries to argue with the law-breaking driver

[2] Ask students: *Why can the brakes stop the moving car within such a short distance? Why can someone running not stop him/herself within such a short distance?* [Q1] (Assume that students have some experience being in a car that stops quickly upon depression of the brakes, even if the car is moving at a speed as high as 20 km/h. If they do not have related experiences, students should be able to imagine that a runner at the same speed could not stop him/herself as quickly. These questions are expected to arouse students' curiosity.)

Experiments of friction coefficients (30 min)

[3] Display a news clip that shows a scooter rider's dangerous ride on nonfriction tiles. Ask students: *Why did the scooter slide less on the asphalt road and more in the "dangerous zone"?* [Q2]

[4] Students with different levels of background knowledge may make teaching more challenging. For Q1, students might point out that the treads on tires and shoes have similar functions. They could attribute the fact that fast runners are unable to stop within a short distance to the runner only wearing two shoes (cars have four tires) or that human legs have less strength than a car engine. If so, point out what is lacking in the students' answers and help them to modify their answers. The experiment of friction coefficient measurement with the tablet PCs should be used once the students are able to offer good answers to Q1

[5] If the students cannot answer Q1, move to Q2. Students are assumed to be able to figure out the question regarding tiles. Make a connection between the tiles and the friction coefficient measurement experiment for the students. (Note: This section follows up with presentations of the definitions of static friction, dynamic friction, and friction coefficient. After acquiring definitions of terms related to friction, students will be told in the next course to use their tablet PCs to measure friction coefficients. They will also be told to bring small-sized items [i.e., palm sized] of different materials.)

[6] Display the video about a racing scooter making a curve (resource 4). Guide students to wonder: *Why can a racing scooter lean on its wheels, but a regular scooter cannot?* [Q3] (It is common to see dangerous riding behaviors like biking without using hands to hold the handlebars, carrying objects that are bigger or longer than the motor, or riding scooters with two feet standing on the pedals. Why are there no cases of riders' leaning on their regular scooters?)

[7] Guide students to conduct the experiment of measuring friction with their tablet PCs. Start the APP called tangent MU1. Place an item (clothes or lightweight strips) on the top of the tablet PC. Then try to place items of different shapes (e.g., rectangle logs, little balls) on the tabletop and see at what angle the tablet tilts to make the item on the tabletop begin to slide

[8] Put 2–3 students in a group (small groups make discussions in experiments easier). For testing bigger items that are not easily measured on a tablet (e.g., wood chunks), one person from the group can hold the item, making the desired angle between the chunk and the floor, and place the tablet on the chunk. The other person can place the item to be measured on the tablet. They can set different angles between the item and the floor by lifting or lowering it. Friction coefficients can be calculated by pressing the function key marked "Setting" in the APP. Students can keep records of the coefficients by trying different items and angles

(continued)

[9] Each group can measure different items, and they can then obtain an average coefficient from measuring the same item 3–5 times. All of the coefficients should be displayed on the blackboard for students to use in making comparisons. Students should here be asked again: *Why can the racing motorbikes lean on their wheels but not the regular ones?* [Q4] Hint to the students to refer back to the data; they will be expected to be able to point out different friction coefficients between racing tires and regular tires and find higher coefficients on the racing tires. (Some students may point out that it's the surface of the floor that makes the friction coefficients different. The teacher should remind students that the racing tracks and regular roads are all paved with asphalt, and paving techniques probably don't contribute to any major differences.)

[10] We can top-glue wood chunks and make an experiment for the students if they are interested in racing tires. The teacher can explain that the racing tires are slick tires, and the tires will heat up and melt if the racing motor passes at a high speed. That would make the tire surface seem like the glue topping

[11] Use PPT to synthesize the main points for students

References

Abbitt, J. T. (2011). Measuring technological pedagogical content knowledge in preservice teacher education: A review of current methods and instruments. *Journal of Research on Technology in Education, 43*(4), 281–300.

Angeli, C., & Valanides, N. (2005). Preservice teachers as ICT designers: An instructional design model based on an expanded view of pedagogical content knowledge. *Journal of Computer-Assisted Learning, 21*(4), 292–302.

Angeli, C., & Valanides, N. (2009). Epistemological and methodological issues for the conceptualization, development, and assessment of ICT-TPCK: Advances in technological pedagogical content knowledge (TPCK). *Computers & Education, 55*(1), 154–168.

Bell, N. (2007). Microteaching: What is it that is going on here? *Linguistics and Education, 18,* 24–40.

Bernhard, J. (2003). Physics learning and microcomputer based laboratory (MBL): Learning effects of using MBL as a technological and as a cognitive tool. In D. Psillos, P. Kariotoglou, V. Tselfes, G. Fassoulopoulos, E. Hatzikraniotis, & M. Kallery (Eds.), *Science education research in the knowledge based society* (pp. 313–321). Dordrecht, The Netherlands: Kluwer.

Boyd, D. J., Grossman, P. L., Lankford, H., Loeb, S., & Wyckoff, J. (2009). Teacher preparation and student achievement. *Educational Evaluation and Policy Analysis, 31*(4), 416–440.

Breslyn, W. J., & McGinnis, R. (2011). A comparison of exemplary biology, chemistry, earth science, and physics teachers' conceptions and enactment of inquiry. *Science Education, 96*(1), 48–77.

Cox, S., & Graham, C. R. (2009). Using an elaborated model of the TPACK framework to analyze and depict teacher knowledge. *TechTrends, 53*(5), 60–69.

Gess-Newsome, J. (1999). Pedagogical content knowledge: An introduction and orientation. In J. Gess-Newsome & N. G. Lederman (Eds.), *Examining pedagogical content knowledge* (pp. 3–17). Dordrecht, The Netherlands: Kluwer.

Gess-Newsome, J., & Lederman, N. G. (1993). Preservice biology teachers' knowledge structures as a function of professional teacher education: A year-long assessment. *Science Education, 77*(1), 25–45.

Grossman, P., & McDonald, M. (2008). Back to the future: Directions for research in teaching and teacher education. *American Educational Research Journal, 45*(1), 184–205.

Harris, J., Grandgenett, N., & Hofer, M. (2010). Testing a TPACK-based technology integration assessment rubric. In D. Gibson & B. Dodge (Eds.), *Proceedings of society for information technology & teacher education international conference 2010* (pp. 3833–3840). Chesapeake, VA: AACE.

Hatfield, R. C. (1989). *Developing a procedural model for the practice of microteaching*. Retrieved from ERIC database. (ED313340)

Jaipal, K., & Figg, C. (2010). Unpacking the "Total PACKage": Emergent TPACK characteristics from a study of preservice teachers teaching with technology. *Journal of Technology and Teacher Education, 18*(3), 415–441.

Jonassen, D., Howland, J., Marra, R., & Crismond, D. (2008). *Meaningful learning with technology* (3rd ed.). Upper Saddle River, NJ: Pearson.

Lee, K., Chai, C. S., & Koh, J. H. L. (2012). Fostering pre-service teachers' TPACK towards student-centered pedagogy. In P. Resta (Ed.), *Society for information technology & teacher education international conference* (pp. 3915–3921). Chesapeake, VA: AACE.

Magnusson, S., Krajcik, J., & Borko, H. (1999). Nature, sources, and development of pedagogical content knowledge for science teaching. In J. Gess-Newsome & N. G. Lederman (Eds.), *Examining pedagogical content knowledge: The construct and its implications for science education* (pp. 95–132). Dordrecht, The Netherlands: Kluwer.

Metcalf, K. (1993). Critical factors in on-campus clinical experiences: Perceptions of preservice teachers. *Teaching Education, 5*(2), 163–174.

Mishra, P., & Koehler, M. J. (2006). Technological pedagogical content knowledge: A framework for teacher knowledge. *Teachers College Record, 108*(6), 1017–1054.

Mitchem, K., Wells, D. L., & Wells, J. G. (2003). Effective integration of instructional technological evaluating professional development and instructional change. *Journal of Technology and Teacher Education, 11*(3), 399–416.

Moersch, C. (1995). Levels of technology implementation (LoTi): A framework for measuring classroom technology use. *Learning and Leading with Technology, 23*(3), 40–42.

Mouza, C. (2003). Learning to teach with new technology: Implications for professional development. *Journal of Research on Technology in Education, 35*(2), 272–289.

Mulhall, P., & Gunstone, R. (2008). Views about physics held by physics teachers with differing approaches to teaching physics. *Research in Science Education, 38*(4), 435–462.

Niess, M. L. (2005). Preparing teachers to teach science and mathematics with technology: Developing a technology pedagogical content knowledge. *Teaching and Teacher Education, 21*(5), 509–523.

Niess, M. L. (2012). Teacher knowledge for teaching with technology: A TPACK lens. In R. Ronau, C. Rakes, & M. L. Niess (Eds.), *Educational technology, teacher knowledge, and classroom impact: A research handbook on frameworks and approaches* (pp. 1–15). Hershey, PA: IGI Global.

Niess, M. L., Ronau, R. N., Shafer, K. G., Driskell, S. O., Harper, S. R., Johnston, C., et al. (2009). Mathematics teacher TPACK standards and development model. *Contemporary Issues in Technology and Teacher Education, 9*(1), 4–24.

Orlich, D. C., Harder, R. J., Callahan, R. C., Kauchak, D. P., Pendergrass, R. A., Keogh, A. J., et al. (1990). *Teaching strategies: A guide to better instruction* (3rd ed.). Lexington, KY: Heath.

Pauline, R. (1993). Microteaching: An integral part of a science methods class. *Journal of Science Teacher Education, 4*(1), 9–17.

Perkins, K., Adams, W., Dubson, M., Finkelstein, N., Reid, S., Wieman, C., et al. (2006). PhET: Interactive simulations for teaching and learning physics. *The Physics Teacher, 44*(18), 18–23.

Pineau, E. L. (1994). Teaching is performance: Reconceptualizing a problematic metaphor. *American Educational Research Journal, 31*(1), 3–25.

Russell, A. L. (1995). Stages in learning new technology: Naïve adult email users. *Computers & Education, 25*(4), 173–178.

Sandholtz, J. H., Ringstaff, C., & Dwyer, D. C. (1997). *Teaching with technology: Creating student-centered classrooms*. New York: Teachers College Press.

Shulman, L. S. (1986). Those who understand: Knowledge growth in teaching. *Educational Researcher, 15*(2), 4–14.

Shulman, L. S. (1987). Knowledge and teaching: Foundations of the new reform. *Harvard Educational Review, 57*(1), 1–22.

van Driel, J. H., Verloop, N., & de Vos, W. (1998). Developing science teachers' pedagogical content knowledge. *Journal of Research in Science Teaching, 35*(6), 673–695.

Yeh, Y.-F., Hsu, Y.-S., Wu, H.-K., Hwang, F.-K., & Lin, T.-C. (2014). Developing and validating technological pedagogical content knowledge-practical (TPACK-practical) through the Delphi survey technique. *British Journal of Educational Technology, 45*(4), 707–722.

Chapter 5
Applying TPACK-P to a Teacher Education Program

Yi-Fen Yeh, Fu-Kwun Hwang, and Ying-Shao Hsu

We propose a teacher community called the learning module design team (LMDT) in which preservice teachers, in-service teachers, and science education researchers work together to enhance each other's TPACK-Practical (TPACK-P). Within the teacher community, in-service teachers designed physics learning applications (APPs) and learning modules with their TPACK-P; preservice teachers then tested the APPs and implemented them into their microteaching. Designing these APPs and learning modules allow in-service teachers in the community to refine their TPACK-P, while implementing these artifacts develops preservice teachers' TPACK-P. A professor who was also a physics teacher educator and science education researcher played the role of a facilitator, ensuring within- and between-group communication. Besides elaborating upon each other's TPACK-P, the LMDT developed a total of 12 android APPs on multitouch tablets to help students better understand physics concepts such as spring resonance, slingshot physics, and friction. This chapter presents the design principles, functions, and features of the 12 APPs; it also describes how these teachers collaborated with each other within the community.

5.1 Introduction

The pursuit of technocentric class instruction has been popular since the last decade of the twentieth century; recently, more focus has been placed on digitizing the

Y.-F. Yeh (✉)
Science Education Center, National Taiwan Normal University, Taipei, Taiwan
e-mail: yyf521@ntnu.edu.tw

F.-K. Hwang
Physics Department, National Taiwan Normal University, Taipei, Taiwan

Y.-S. Hsu
Graduate Institute of Science Education, National Taiwan Normal University, Taipei, Taiwan

© Springer Science+Business Media Singapore 2015
Y.-S. Hsu (ed.), *Development of Science Teachers' TPACK*,
DOI 10.1007/978-981-287-441-2_5

content of teaching and learning as well as on teacher quality. In traditional physics classrooms, teachers often use analogy as an instructional strategy to help students imagine and comprehend abstract concepts. Conceptualizing the atom as similar to the solar system is one example (Harrison & Treagust, 1993; Podolefsky & Finkelstein, 2006). Nowadays, with help from computers, multimedia applications can effectively present micro- or macrophenomena in science and make learning interactive and individualized (Ainsworth, 2006; Mayer, 1999; Plass, Chun, Mayer, & Leutner, 1998; Teasley & Rochelle, 1993; Wu, Krajcik, & Soloway, 2001). The unobservable, abstract concepts (i.e., waves, projectile motion) or imaginary analogies can be visually concretized or idealized through simulation, thereby lowering the cognitive demands on students (de Jong & van Joolingen, 1998; Goldstone & Son, 2005). Selecting and using appropriate technology to make context instruction more comprehensible requires teachers to develop their technological pedagogical content knowledge (TPACK; Mishra & Koehler, 2006).

Curriculum digitization is not the same as curriculum computerization. Pedagogical concerns and technological affordances need to be considered. Experienced teachers know what their students need in learning, as well as what they need to make their teaching more comprehensible. Therefore, in this chapter, we propose a teacher community that develops and reinforces the bidirectional development of both experienced teachers' and preservice teachers' TPACK in teaching practices (TPACK-Practical). Experienced teachers designed simulation-based applications (APPs) and learning modules for preservice teachers to use; the feedback from the users (preservice teachers and students) was then directed back to the designers (in-service teachers), offering them information to use when improving their APPs and modules as well as elaborating their TPACK-P. The TPACK-P of teachers with different proficiency levels was believed to be elaborated through the tasks of designing, implementing, and modifying these learning APPs and modules. This chapter also offers the design principles used when developing the learning APPs, as well as the major features of each APP, in order to inform future designers of science learning software.

5.2 Simulations for Science Education

ICT-based interventions can either be used to enhance the practical investigation or as a virtual alternative to real practical work where a simulation supports exploration of the investigative model through a computerized representation of the phenomena under study (McFarlane & Sakellariou, 2002, p. 221).

Science is a subject that demands students explore the natural world, and simulations can offer imitation or operational models and interactive environments to make complex or inaccessible phenomena friendly to users. Research also found simulation a useful tool for constructing students' understanding of concepts, developing their scientific inquiry abilities, and enhancing their science learning motivations (Baxter, 1995; de Jong & van Joolingen, 1998; Eylon, Ronen, & Ganiel 1996;

Goldstone & Son, 2009; Reid, Zhang, & Chen, 2003; Rutten, van Joolingen, & van der Veen, 2012; Zacharia, 2007; Zacharia & Anderson, 2003). Since teachers' attitudes toward and intentions for teaching with simulations influence their choice of appropriate teaching models or activities (Zacharia, 2003), preparing science teachers to be knowledgeable about and competent in using simulations will be fundamental in promoting teachers' uses of simulation-based curricula to assist student science learning.

5.2.1   Simulations in Science Learning

Interest in pursuing simulation applications for science teaching and learning has increased in recent years. The Physics Education Technology (PhET) project, which created and released approximately 50 simulation-based programs for physics teachers worldwide, was one such successful effort focused on physics learning and teaching. These simulations were created for "supporting students in constructing a robust conceptual understanding of the physics through exploration" (Perkins et al., 2006, p. 18). To achieve such a goal, interactive animations or responsive systems are purposefully built to encourage students' self-explorations. Simulations in Java™ or Flash® format offer easy access for students seeking to perform scientific explorations or for teachers looking to demonstrate phenomena in their lectures (Wieman, Adams, Loeblein, & Perkins, 2010). These simulation-based learning tools can be embedded into learning modules. For example, Chang, Chen, Lin, and Sung (2008) constructed a thematic course about optical reflection and refraction in which students were hypertext prompted to activate their prior knowledge and engage in scientific inquiry through the manipulation of built-in simulations (e.g., make, test, and form conclusions about their hypotheses). Web-based inquiry science environment (WISEm 1996–2003) is another example of how simulations can be implemented to assist students' self-directed or group inquiry learning tasks.

5.2.2   Tablet PCs as Good Carriers of Simulations

Tablet PCs (hereafter called tablets) are small, portable computers that are lightweight, reasonably durable, and mobile. Tablets can do most of what home computers do (though often with less advanced functions); however, they offer teachers better control and versatility in displaying content, making impromptu edits, and switching between programs and other applications (Mock, 2004). Built-in sensors (e.g., accelerometers, gyroscopes, and gravity sensors) make tablets sensitive to users' body motions, where the kinetic and haptic experiences involved in learning with built-in sensors reinforce students' mental representations and concept construction (de Koning & Tabbers, 2011; Wang, Wu, Chien, Hwang, & Hsu, 2015). The tablet's multitouch screen, which receives input from single or multiple users simultaneously, encourages efficient and equitable participation among group members during their discussion and collaboration efforts (Marshall, Hornecker, Morris, Dalton, & Rogers, 2008; Piper, O'Brien, Morris, & Winograd, 2006; Rick, Harris, Marshall, Fleck, Yuill, & Rogers, 2009). All these sensors make tablets function like microcomputer-based laboratories (MBLs), which allow students to construct their inquiry ability by engaging them in tasks of data collection and analysis.

5.2.3 Design Principles for Simulation-Based APPs and Learning Modules

Based on previous studies regarding how simulations and tablets can be implemented in teaching and learning, we developed design principles that should be considered when designing simulation-based APPs for physics learning. These include:

Concept Construction Scientific principles, concepts, and facts can be embedded in simulations wherein students can explore and construct conceptual and operational models (de Jong & van Joolingen, 1998).

Model Exploration Programmers set formulas with default values and predefined ranges of parameters in order to allow complex natural phenomena to be examined through variable manipulation activities or virtual experiments (McFarlane & Sakellariou, 2002). Students interact with these model-based learning programs, allowing them to practice and strengthen their scientific thinking (e.g., variable identification or manipulation, hypothesis, or model testing).

Real Data Collection Built-in sensors (e.g., gravity or multitouch sensors) make tablets function much like MBLs. Students can have personalized and rewarding science learning experiences when they are encouraged to actively use MBLs to collect data and make related analyses (Thornton & Sokoloff, 1990).

Format Flexibility Using appropriate pedagogy (e.g., predict–observe–explain [P–O–E], inquiry learning) to support simulation implementation in science classroom contexts facilitates student learning about the nature of science (Monaghan & Clement, 1999).

These design principles indicate that the designers of science learning APPs need to be equipped with professional knowledge not only in the target science topics and programming but also in the pedagogy of teaching and learning the target scientific concepts or practices. Teachers can be good curriculum designers because they are experts in knowledge delivery and student learning progress; but they may lack knowledge and experience with technology-infused curriculum design because, in most cases, they are not programmers (Sandholtz & Reilly, 2004). Professional programmers do not normally make good curriculum designers either because they lack professional knowledge about science teaching and learning. Therefore, to design and develop effective science learning APPs requires a team that is composed of both science teachers and programmers. The more teachers and programmers there are on the design team, the more pedagogically advanced the instructional artifacts are likely to be.

5.3 Development of Teachers' TPACK-P

Simulation-based APPs can be used as tools by teachers to assist their instruction or as curricula by students seeking self-study. To design, develop, and implement these curriculum-driven, technology-assisted learning tools demand not only a profound TPACK but also related teaching experiences. It is not necessary to view the instructional artifacts that teachers produce fully developed end products ready for wide-scale educational use; rather, these technology-infused tasks can be viewed as activities in the professional learning journey for teachers. Considering that TPACK is dynamic and situated in the context and nature of the learning situation (Cox & Graham, 2009; Doering, Veletsianos, Scharber, & Miller, 2009; Mishra & Koehler, 2006), teachers can develop and refine their TPACK through actions in the design–implementation–evaluation–modification cycle.

5.3.1 TPACK for Curriculum Designing

How teaching material is designed or implemented into instruction requires teachers' deliberation on factors that influence student learning of the target learning outcomes (e.g., concepts, affect, skills, practices/processes, etc.). According to Shulman (1986), teachers with well-developed pedagogical content knowledge (PCK) are able to identify:

The most regularly taught topics in one's subject area, the most useful forms of representation of those ideas, the most powerful analogies, illustrations, examples, explanations, and demonstrations including an understanding of what makes the learning of specific concepts easy or difficult, the conceptions and preconceptions that students of different ages and backgrounds bring with them to the learning. (p. 9)

All of these pieces of knowledge fulfill the qualification requirements for effective teachers of disciplinary content. As teachers gain more experiences in teaching, teachers' PCK is transformed into *craft knowledge*, which refers to the personally constructed wisdom that teachers rely on to facilitate their instruction. In that sense, we can assume that experienced teachers' PCK serves as both a benchmark and an objective for preservice and/or novice teachers.

Technology-rich learning and teaching environments are an additional component to contemporary PCK. Therefore, as technology is more and more often considered and added to the framework of PCK, expanding teachers' knowledge of teaching with technology—their TPACK—becomes a necessary capability for contemporary teachers (Mishra & Koehler, 2006). Situated mapping of technological affordances, representations, pedagogy, and learners for different subject topics is important to further transform teachers' TPACK into an active knowledge that embraces student learning (Angeli & Valanides, 2009). Tsai and Chai (2012) proposed the term *design thinking* to describe the essence of teachers' TPACK and defined it as convergent knowledge for teachers to use in creating teaching practices where technological advances are meaningfully applied to support teachers' curriculum design and implementation.

5.3.2   TPACK-P Through a Teacher Community

Designing simulation-based APPs for science teaching and learning demands requires that teachers start with comprehensive considerations of content, learners, and pedagogy within the specified disciplines, topics, and learning outcomes. Activities like curriculum designing and instructional enactment allow designer teachers and practitioner teachers to develop and further refine their TPACK. We assume such TPACK is later transformed into TPACK-P when they design and implement simulation-based APPs to assist their science instruction through stages of design and planning, enactment, and assessment (see Chapter 2; Yeh, Hsu, Wu, Hwang, & Lin, 2014). Given that TPACK-P is experience based and personally constructed, teachers within the teacher community with different levels of TPACK-P can receive different levels of assistance from community members in an ever-changing mentoring dynamic.

"Teachers helping teachers" is the main reason why teacher communities are a necessity (Feiman-Nemser, 2001, p. 1043). All practicing teachers in the community support each other when solving instructional problems and planning for their solutions' sustainability. Mentors rationalize their interwoven considerations when mapping technological affordances to specific subject-area content (C. Chang, Chien, Chang, & Lin, 2012) or offering explicit support to assist preservice teachers in designing instructional artifacts (Koehler & Mishra, 2005). Their mature pedagogical insights and teaching experiences can be good resources to help preservice teachers address the theory–practice gap; but in fact, these experienced teachers also benefit from interacting with novice teachers and encountering their energy and idealism. Novice teachers in most cases have higher levels of technological literacy than experienced teachers—they are the *Net Generation* and likely bring contemporary information about advanced digital electronics and innovative uses of these technological tools to the educational context and their mentor teachers. Since teachers from various places in their career development have different backgrounds, specializations, and beliefs, a community where teachers can learn from each other and offer different perspectives is constructive to teachers' instructional quality and to their own professional development.

Forming a teacher community where teachers with different TPACK-P proficiency levels and specialties participate offers these teachers a platform to share ideas and learn from each other. Teachers in the community can be learners and professionals at the same time, since they learn from each other (Abdal-Haqq, 1996; Lawless & Pellegrino, 2007; Newmann & Associates, 1996). Self and collaborative reflections among teachers, which we called critical collegiality, further boost teachers' reflecting-on-action and reflecting-in-action, intellectual virtues, and communicative skills (Lord, 1994). Such teacher collaboration can bring positive impacts to the quality and practicality of the instructional outputs as well as to the refinement of teachers' expertise. Other features such as offering longitudinal support and technology access are also factors sustaining teachers' participation willingness and the richness of the community.

5.4 Learning Module Design Team

Teachers' TPACK-P is complex knowledge constructs that are continuously refined by teachers' experience accumulation and regular practice in instructional design and enactment. With the purpose of developing learning APPs that can effectively facilitate physics teachers' instruction, we propose a learning module design team (LMDT) within which teacher educators, experienced teachers, and preservice teachers collaboratively work. Teacher educators and experienced teachers take the lead roles in designing instructional artifacts (e.g., APPs, learning modules), while preservice teachers reflect on their experiences with the APPs and provide user feedback in modifying the artifacts. These authentic engineering design, evaluation, and modification experiences further elaborate the TPACK-P of the teacher educators, experienced teachers, and preservice teachers who participate (Fig. 5.1).

5.4.1 APP and Learning Module Designers

The LMDT consisted of a teacher educator, four experienced physics teachers, and 11 preservice teachers who collaborated with each other on developing simulation-based physics learning modules. The professor had been educating physics teachers

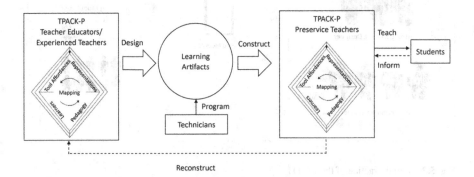

Fig. 5.1 Rationale of LMDT (Modified from Angeli & Valanides, 2009)

for over 20 years, while the experienced teachers had 19 years in teaching on average (43, 18, 10, and 5 years). These experienced teachers had collaborated with the
professor on designing and evaluating physics curricula for an average of 9 years
(13, 13, 6, 4 years). Regular group meetings were held for learning the APP and the
learning module design and modifications. Figure 5.2 shows how the teacher community operated in actual practice.

Within the LMDT, four experienced physics teachers interested in APP development were responsible for the APP design. Among them, one teacher played the
role of lead programmer. The discussions among the experienced teachers in the
APP design meetings covered topics such as (a) what affordances of tablets could
be useful to student learning, (b) what physics concepts students found difficult to
understand and whether those concepts could be presented through APPs on tablets,
(c) what learning objectives for each concept needed to be achieved, (d) how concretized or idealized concepts could be achieved and presented, and (e) the possible
models that would activate the mechanism of the target concepts. They endeavored
to seek a balanced mapping of content representations, the unique affordances of
tablets, learners' cognitive development, and pedagogy for each APP and concept.
For example, they believed that the built-in gravity sensors in tablets could be effective tools for students measuring the components of gravitational field strength and
help them to physically sense the gravity and visualize how projectiles move on a
trajectory. Therefore, they included principles of independent motion so that students could conceptualize how motion is influenced by its initial speed and gravity
through manipulating and interacting with the APPs.

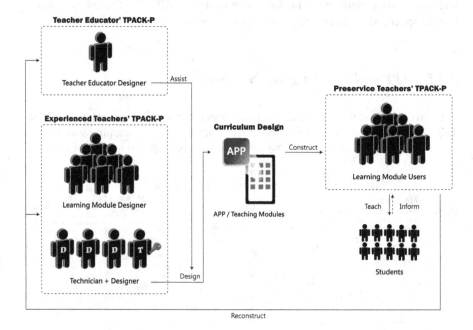

Fig. 5.2 Actual practice of the LMDT

The prototypes of these APPs were later sent to the learning module designers who were experienced teachers and members of the larger teacher community (though not involved in the APPs' design). These designers were encouraged to familiarize themselves with these APPs, comment on how to make these APPs more friendly and meaningful to their users, and generate learning modules for use when teaching with the associated APPs. Their user feedback was collected and considered by the designers when improving the learning APPs. When developing the learning modules, these module designers shared how they taught related concepts and how their students might react to the use of the APP. Through several discussions, they all agreed that the P–O–E approach would be a useful strategy for guiding students' learning of projectile motion while using the simulation-based APPs. Finally, these experienced teachers developed learning modules with guiding questions and suggested steps for teachers to use in helping students acquire the concepts of projectile motion and scientific thinking.

These two groups of experienced teachers (the APP designers and learning module designers) refined their TPACK-P as they produced, reflected upon, and evaluated the APPs and corresponding learning modules. Their TPACK-P was further shaped by their negotiating of an array of factors critical to instructional artifact design, such as what scientific models needed to be embedded, how the scientific phenomena were presented, who the target students were, and where these APPs and learning modules could be used. It is the engineering design process (e.g., invent, evaluate, revise) that repeatedly requires teachers to engage best solutions and then develop beta versions of the APPs, which science teachers can then use to assist their science teaching.

5.4.2 APP and Learning Module Practitioners

Preservice teachers comprised the other part of the LMDT; they played dual roles of artifact practitioners implementing the APPs and of outsider APP testers. They represented and offered user feedback from the perspective of teachers not experienced in teaching with technology or using such APPs and learning modules. Their personal use and implementation experiences in microteaching and teaching internships provided a fresh perspective for the APP designers.

Learning from exemplar teachers and through actual teaching practices is a direct way of developing preservice teachers' TPACK-P. The professor of the LMDT, who was also the instructor of the physics teaching practicum, provided each preservice teacher with a tablet when the course began in order to familiarize them with tablet manipulation. He rationalized how the APPs and learning modules were designed and demonstrated how these APPs and modules could assist them in their future physics instruction. While implementing the APPs in their teaching, these preservice teachers were able to generate some innovative instructional uses; they also brought in feedback from high school students responding to the use of these APPs. These innovative uses and user feedback were later collected for the APP and learning module designers to use in improving the original APPs and modules. Newer versions of the APPs were created to better serve both teaching and learning needs as well as to address different educational purposes. For example, the original Spring Resonance APP only allowed users to drag and use one mass in each trial, but the newer version allowed users to drag and use up to three different masses in trial tests. Displaying the resonance of three springs with different masses together was expected to help students better visualize the resonance with comparisons and under controlled conditions.

5.4.3   Examples of the APPs

The physics APPs and learning modules have repeatedly been improved and expanded to better accommodate teachers' and students' needs, based on feedback from teachers who used the APPs within the LMDT, from teachers participating in other professional development workshops, and from high school students who used the APPs as part of their science classes. The descriptions, functions, and distinctive features of the 12 APPs are listed in Table 5.1.

5.5   Final Remarks

Physics APPs and learning modules were the main products of the "Aim for the Top University" project. This project was missioned with finding ways to prepare a friendly science learning environment and quality science teachers for the digital era and to equip students with good science concepts and scientific thinking. The underlying mechanism for running a teacher community (i.e., the LMDT) involves valuing and allocating teachers' wisdom and heterogeneous TPACK-P proficiencies through an engineering design cycle of invention–trial–feedback–redesign of instructional artifacts. Participating teachers were expected to develop and refine their TPACK-P from their tangential involvement not only in the APP and learning module development but also through their collaborations with other teachers. Iterative and multidirectional experiences transformed preservice and in-service teachers' TPACK-P through authentic collaborative practice for the mutual benefit of all.

We also established and presented the rationales for and design features of simulation-based APPs for science teaching and learning. These simulation-based learning APPs were designed to allow students to explore abstract physics or unobservable scientific phenomena by engaging them in simultaneous haptic manipulation of multiple variables. Based on an empirical study that we did for knowing the effectiveness of the APPs we developed, high school students showed improvements in their understanding of projectile motion and collision after taking the module-based course where related APPs and the strategy of P–O–E were implemented (Wang et al., 2015). Considering that simulation-based learning APPs on mobile devices can make science teaching and learning less effortful, it is worthwhile

Table 5.1 Features of the physics APPs

Physics learning APP	Description	Multitouch[a]	Built-in sensors[b]	Kinetic manipulation[c]	Experimentation[d]	Variable manipulation[e]
Spring resonance	Three mass–spring systems with different nature frequencies were placed on the tablet screen. Each system consisted of a mass with a spring on both ends attached to a wall. Users can move the mass up and down by tilting the tablet to a different angle. They are expected to find the right frequency of tilting motion so that one of the springs is in resonance with the motion.	X	X	X	X	X
Slingshot physics	Two slingshots with different spring constants (k1, k2, k2 = 2*k1) are built in the APP for users to predict which slingshot would shoot their projectile as far as possible under the condition that the maximum force stretching slingshots each time remains the same. Users can explore how the tension of the slingshot influences the distance traveled.	X			X	X

(continued)

Table 5.1 (continued)

Physics learning APP	Description	Multitouch[a]	Built-in sensors[b]	Kinetic manipulation[c]	Experimentation[d]	Variable manipulation[e]
Friction	Static friction coefficients can be calculated in terms of the maximum angle that can be reached before one of the items begins to slide. Users can place an object on the top of the tablet and find the friction coefficient from the built-in angle-measuring program by raising one side of the tablet. This APP allows users to adjust frictions (μ_s, μ_k) and see the vectors of force and acceleration from the arrows displayed on the screen to help understand the force reactions from acceleration, gravity, and friction.	X	X	X	X	X
Projectile motion	Users can finger press the tablet screen to see projectiles and set the initial speed of the projectile by dragging their fingers. After releasing their fingers, these projectiles will move due to the physics laws. Users can see how the magnitude and direction of gravity influence the motion of projectiles by tilting the tablet where the built-in gravity sensor can detect the angles between the tablet and the surface. Trajectories of the projectiles can be displayed, helping users to visualize how different variables (e.g., speed, gravity) in physics laws contribute to the projectile motion. Similar concepts: Curtain Throw		X	X	X	

Collision	Users can explore what factors contribute to the collision of two objects by changing the velocity, the direction of motion, etc. The APP allows users to set up the initial conditions of the projectiles (e.g., direction, speed) by touching and applying pressure on the projectiles. Pressing the "Play" button on the top-right corner starts the motion of the two objects with preset conditions. Users can display the trajectories and the vectors involved in the motion for better visualizing and understanding how gravity and velocity interact onto the moving objects.		X	X	X	X
Graphs of motion 1-D	The APP is built-in with some predefined V–t graphs and offers blank graphs for users to draw X–t and/or a–t graphs for different V–t graphs. In order to let users visualize possible interactions of their predictions, the APP can display the corresponding V–t graph from the user's X–t and/or a–t graphs (with different colors). Similar concepts: Graphs of Motion 2-D		X			
Visual weight	This APP uses the built-in sensors of the tablet (i.e., gyroscope and gravitational sensors) to allow users to see how gravity influences the visual weight of the objects. By tilting or flipping the tablet, users can observe how the visual weights correspondingly change according to the varying acceleration.		X	X	X	

(continued)

Table 5.1 (continued)

Physics learning APP	Description	Multitouch[a]	Built-in sensors[b]	Kinetic manipulation[c]	Experimentation[d]	Variable manipulation[e]
Pendulum	Users can observe the oscillation of the pendulum by dragging the pendulum drop, adjusting the mass, and changing the gravitational acceleration by tilting the tablet. Students can calculate its frequency period based on the data collected from the built-in timer. Variables like angles, length of the drop, and mass of the pendent can be changed. They can also observe and discuss the simple harmonic motion of the spring–mass particles.		X	X	X	X
Find the center of mass	This APP uses the built-in sensors (i.e., gyroscope, gravitational sensors) to familiarize students with the approach that physicists use to find the center of mass for objects. Users can flip the tablet around to find the center of mass for the tablet.		X	X	X	

Dot to motion	Users can draw the motion trajectory and make explanations in practices with different conditions by setting moving speeds at 2 m/s in gravitational fields. The mesh in the background helps users find where the dots are at the x and y axes.			X	

Note

[a]Multitouch: Enables the function of more than two simultaneous touch points

[b]Built-in sensors: Enable the built-in sensors of tablets (e.g., accelerometers, gyroscopes, and gravity sensors) to help students experience science phenomena or collect related data

[c]Kinetic manipulation: Engages students to physically interact with tablets to understand abstract concepts. For example, students can see and feel weightlessness in the Visual Weight APP when the tablet is thrown into free-fall motion

[d]Experimentation: Simplifies the context and tries to construct an ideal environment for students to do experiments

[e]Variable manipulation: Allows users to manipulate variables and see the results

for teacher educators to initiate an engineering design cycle for building an authentic and effective teacher community where they can benefit from mutual support or to customize curricula together based on their needs. All these endeavors will ultimately lead to the enhancement of student learning.

References

Abdal-Haqq, I. (1996). *Making time for teacher professional development*. Retrieved from ERIC database. (ED400259).

Ainsworth, S. (2006). DeFT: A conceptual framework for considering learning with multiple representations. *Learning and Instruction, 16*(3), 183–198.

Angeli, C., & Valanides, N. (2009). Epistemological and methodological issues for the conceptualization, development, and assessment of ICT-TPCK: Advances in technological pedagogical content knowledge (TPCK). *Computers & Education, 55*(1), 154–168.

Baxter, G. P. (1995). Using computer simulations to assess hands-on science learning. *Journal of Science Education and Technology, 4*(1), 21–27.

Chang, C.-Y., Chien, Y.-T., Chang, Y.-H., & Lin, C.-Y. (2012). MAGDAIRE: A model to foster pre-service teachers' ability in integrating ICT and teaching in Taiwan. *Australasian Journal of Educational Technology, 28*(6), 983–999.

Chang, K. E., Chen, Y. L., Lin, H. Y., & Sung, Y. T. (2008). Effects of learning support in simulation-based physics learning. *Computers & Education, 51*(4), 1486–1498.

Cox, S., & Graham, C. R. (2009). Using an elaborated model of the TPACK framework to analyze and depict teacher knowledge. *TechTrends, 53*(5), 60–69.

de Jong, T., & van Joolingen, W. R. (1998). Scientific discovery learning with computer simulations of conceptual domains. *Review of Educational Research, 68*(2), 179–201.

de Koning, B. B., & Tabbers, H. K. (2011). Facilitating understanding of movements in dynamic visualizations: An embodied perspective. *Educational Psychology Review, 23*(4), 501–521.

Doering, A., Veletsianos, G., Scharber, C., & Miller, C. (2009). Using the technological, pedagogical, and content knowledge framework to design online learning environments and professional development. *Journal of Educational Computing Research, 41*(3), 319–346.

Eylon, B., Ronen, M., & Ganiel, U. (1996). Computer simulations as a tool for teaching and learning: Using a simulation environment in optics. *Journal of Science Education and Technology, 5*(2), 93–110.

Feiman-Nemser, S. (2001). From preparation to practice: Designing a continuum to strengthen and sustain teaching. *Teachers College Record, 103*(6), 1013–1055.

Goldstone, R. L., & Son, J. Y. (2005). The transfer of scientific principles using concrete and idealized simulations. *Journal of the Learning Sciences, 14*(1), 69–110.

Harrison, A. G., & Treagust, D. F. (1993). Teaching with analogies: A case study in grade 10 optics. *Journal of Research in Science Teaching, 30*(10), 1291–1307.

Koehler, M. J., & Mishra, P. (2005). What happens when teachers design educational technology? The development of technological pedagogical content knowledge. *Journal of Educational Computing Research, 32*(2), 131–152.

Lawless, K. A., & Pellegrino, J. W. (2007). Professional development in integrating technology into teaching and learning: Knowns, unknowns, and ways to pursue better questions and answers. *Review of Educational Research, 77*(4), 575–614.

Lord, B. (1994). Teachers' professional development: Critical colleagueship and the role of professional communities. In N. Cobb (Ed.), *The future of education: Perspectives on national standards in education* (pp. 175–204). New York, NY: College Entrance Examination Board.

Marshall, P., Hornecker, E., Morris, R., Dalton, N. S., & Rogers, Y. (2008). When the fingers do the talking: A study of group participation with varying constraints to a tabletop interface. *Proceedings of the of IEEE Tabletops and Interactive Surfaces* (pp. 33–40) (Tabletop '08).

Mayer, R. E. (1999). *The promise of educational psychology: Learning in the content areas.* Upper Saddle River, NJ: Prentice Hall.

McFarlane, A., & Sakellariou, S. (2002). The role of ICT in science education. *Cambridge Journal of Education, 32*(2), 219–232.

Mishra, P., & Koehler, M. (2006). Technological pedagogical content knowledge: A framework for teacher knowledge. *Teachers College Record, 108*(6), 1017–1054.

Mock, K. (2004). Teaching with tablet PC's. *Journal of Computing Sciences in Colleges, 20*(2), 17–27.

Monaghan, J. M., & Clement, J. (1999). Use of a computer simulation to develop mental simulations for understanding relative motion concepts. *International Journal of Science Education, 21*(9), 921–944.

Newmann, F. M., & Associates. (1996). *Authentic achievement: Restructuring schools for intellectual quality.* San Francisco, CA: Jossey-Bass.

Perkins, K., Adams, W., Dubson, M., Finkelstein, N., Reid, S., & Wieman, C., & LeMaster, R. (2006). PhET: Interactive simulations for teaching and learning physics. *The Physics Teacher, 44*(18), 18–23.

Piper, A. M., O'Brien, E., Morris, M. R., & Winograd, T. (2006). SIDES: A cooperative tabletop computer game for social skills development. In *Proceedings of the 2006 20th Anniversary Conference on Computer-supported Cooperative Work.* Banff, Canada: Association for Computing Machinery.

Plass, J. L., Chun, D. M., Mayer, R. E., & Leutner, D. (1998). Supporting visual and verbal learning preferences in a second-language multimedia learning environment. *Journal of Educational Psychology, 90*(1), 25–36.

Podolefsky, N. S., & Finkelstein, N. D. (2006). Use of analogy in learning physics: The role of representations. *Physical Review Special Topics – Physics Education Research, 2*(2), 020101-1-10.

Reid, D. J., Zhang, J., & Chen, Q. (2003). Supporting scientific discovery learning in a simulation environment. *Journal of Computer Assisted Learning, 19*(1), 9–20.

Rick, J., Harris, A., Marshall, P., Fleck, R., Yuill, N., & Rogers, Y. (2009). Children designing together on a multi-touch tabletop: An analysis of spatial orientation and user interactions. *Proceedings of the 7th international conference on interaction design and children* (pp. 106–114). New York, NY: Association for Computing Machinery.

Rutten, N., van Joolingen, W. R., & van der Veen, J. T. (2012). The learning effects of computer simulations in science education. *Computers & Education, 58*(1), 136–153.

Sandholtz, J., & Reilly, B. (2004). Teachers, not technicians: Rethinking technical expectations for teachers. *Teachers College Record, 06*(3), 487–512.

Shulman, L. S. (1986). Those who understand: Knowledge growth in teaching. *Educational Researcher, 15*(2), 4–14.

Teasley, S. D., & Rochelle, J. (1993). Constructing a joint problem space: The computer as a tool for sharing knowledge. In S. P. Lajoie & S. J. Derry (Eds.), *Computers as cognitive tools* (pp. 229–258). Hillsdale, NJ: Lawrence Erlbaum.

Thornton, R. K., & Sokoloff, D. R. (1990). Learning motion concepts using real-time microcomputer-based laboratory tools. *American Journal of Physics, 58*(9), 858–867.

Tsai, C.-C., & Chai, C. S. (2012). The "third"-order barrier for technology-integration instruction: Implications for teacher education. *Australasian Journal of Educational Technology, 28*(6), 1057–1060.

Wang, J.-Y., Wu, H.-K., Chien, S.-P., Hwang, F.-K., & Hsu, Y.-S. (2015). Designing applications for science learning: Facilitating high school students' conceptual understanding by using tablet PCs. *Journal of Educational Computing Research, 51*(4), 441–458.

Wieman, C. E., Adams, W. K., Loeblein, P., & Perkins, K. K. (2010). Teaching physics using PhET simulations. *The Physics Teacher, 48*(4), 225–227.

WISE – v4 Web-Based Inquiry Science Environment. (1996–2013). *Homepage.* Retrieved from http://wise.berkeley.edu/webapp/index.html

Wu, H.-K., Krajcik, J., & Soloway, E. (2001). Promoting understanding of chemical representations: Students' use of a visualization tool in the classroom. *Journal of Research in Science Teaching, 38*(7), 821–842.

Yeh, Y.-F., Hsu, Y.-S., Wu, H.-K., Hwang, F.-K., & Lin, T.-C. (2014). Developing and validating technological pedagogical content knowledge-practical (TPACK-practical) through the Delphi survey technique. *British Journal of Educational Technology, 45*(4), 707–722.

Zacharia, Z. (2003). Beliefs, attitudes, and intentions of science teachers regarding the educational use of computer simulations and inquiry-based experiments in physics. *Journal of Research in Science Teaching, 40*(8), 792–823.

Zacharia, Z. C. (2007). Comparing and combining real and virtual experimentation: An effort to enhance students' conceptual understanding of electric circuits. *Journal of Computer Assisted Learning, 23*(2), 120–132.

Zacharia, Z., & Anderson, O. R. (2003). The effects of an interactive computer-based simulation prior to performing a laboratory inquiry-based experiment on students' conceptual understanding of physics. *American Journal of Physics, 71*(6), 618–629.

Part III
The Integrative Model of TPACK

Chapter 6
Developing Preservice Teachers' Sensitivity to the Interplay Between Subject Matter, Pedagogy, and ICTs

Yu-Ta Chien and Chun-Yen Chang

Mishra and his colleagues' notion of technological pedagogical content knowledge (TPCK, renamed as TPACK in Thompson & Mishra, 2007–2008) theorizes that the required knowledge for teachers to teach with information and communication technology (ICT) involves comprehensive understanding of the transactional interplay between the subject matter being taught, the pedagogy being used, and the ICT tools being adopted in teaching practice. Aligning with the conceptualization of TPACK, developing preservice teachers' sensitivity to the interplay between subject matter, pedagogy, and ICT is a key objective for teacher preparation programs. Based on the theoretical framework of cognitive apprenticeship, we propose a 4-phase cyclic MAGDAIRE model (abbreviated from modeled analysis, guided development, articulated implementation, and reflected evaluation) to develop preservice teachers' sensitivity to the interplay between subject matter, pedagogy, and ICT. MAGDAIRE is subsequently employed to enhance the science teacher education courses of National Taiwan Normal University. The TPACK conceptual framework is adapted as an analytic tool to examine the growth in preservice science teachers' knowledge about technology integration in teaching. The results of the studies and courses indicate that, within MAGDAIRE, these preservice science teachers' reasoning on the use of ICT transited toward a more connected model in which ICT is jointly considered with subject matter and/or pedagogy. Moreover, these preservice teachers' development of TPACK stimulated them to modify their practice. In this chapter, the details of MAGDAIRE and a synthesis of the studies into MAGDAIRE are reported.

Y.-T. Chien
Science Education Center, Graduate Institute of Science Education,
National Taiwan Normal University, Taipei, Taiwan

C.-Y. Chang (✉)
Science Education Center, Graduate Institute of Science Education,
Department of Earth Sciences, National Taiwan Normal University, Taipei, Taiwan
e-mail: changcy@ntnu.edu.tw

© Springer Science+Business Media Singapore 2015
Y.-S. Hsu (ed.), *Development of Science Teachers' TPACK*,
DOI 10.1007/978-981-287-441-2_6

6.1 Introduction

In reviewing the education policies issued by the Taiwan Ministry of Education (MOE), one can find that the teaching profession in contemporary Taiwanese society is expected to adopt a new element: able to teach with information and communication technology (ICT). As stated in the Information Literacy Competence Standards for Elementary and Junior High School Teachers (MOE, 2001), teachers need to know how best to make use of a range of ICT to support teaching and learning. Taking a global perspective, numerous national-level movements have been engaged to infuse ICT into classrooms in the regions of Europe (Pelgrum & Doornekamp, 2009), North America (International Society of Technology in Education [ISTE], 2008), and Asia-Pacific (Organization for Economic Cooperation and Development [OECD], 2008). In Taiwan, the MOE heavily invested resources into reducing the student-to-computer ratio with Internet access for elementary and junior high school schools; generally, the availability of ICT should no longer be a problem for teachers. According to the MOE's white paper on Information Technology Education for Elementary and Junior High Schools 2008–2011 (MOE, 2008), it is expected that more than 90 % of elementary and junior high school teachers will enrich their teaching practice with ICT. At the same time, we, as teacher educators, must reflect seriously on whether the current teacher preparation courses are able to meet this demand.

Table 6.1 provides the common structure of teacher preparation programs approved by the MOE. Clearly, the main purpose of these courses is to nourish preservice teachers' sense of what teaching is as well as how to teach. Within this program structure, the Instructional Media course is the one most related to ICT. This course focuses on basic operations of word processing, spreadsheet, and some presentation tools to patch up preservice teachers' ICT skills in a stand-alone format. Without a doubt, being able to personally use ICT is a fundamental factor contributing to teachers' use of ICT in teaching (Govender & Govender, 2009; Mahdizadeh,

Table 6.1 Common structure of teacher preparation programs in Taiwan

Field	Course
Foundations of education	Introduction to education
	Educational psychology
	Sociology of education
	Philosophy of education
Instructional methods	Principles of teaching
	Classroom management
	Educational assessment and evaluation
	Theory and practice in counseling
	Curriculum development and design
	Instructional media
Teaching practicum	Teaching materials and methods
	Practicum

Biemans, & Mulder, 2008; Sørebø, Halvari, Gulli, & Kristiansen, 2009). However, this approach has been widely criticized for its failure to correlate teachers' ICT skills with their teaching practice (Angeli, 2005; ISTE, 2008; Jang & Chen, 2010; Mishra & Koehler, 2006; Niess, 2005; OECD, 2010; Wilson, 2003). Angeli (2005) indicated that this type of ICT course often gives preservice teachers an impression that ICT is the subject matter to be learned rather than an instructional tool. It is commonly perceived by preservice teachers that the letters conventionally written in chalk are now being displayed by the latest presentation tools, but nothing else appears to have changed or needs to be changed; if so, why should teachers bother about the use of ICT in teaching? As a result, preservice teachers may lose sight of how ICT tools can be leveraged to serve educational purposes.

Mishra and his colleagues' notion of technological pedagogical content knowledge (TPCK, renamed as TPACK in Thompson & Mishra, 2007–2008) provides teacher educators with an alternative to rethink what we expect preservice teachers to do with ICT in teacher preparation courses. The conceptualization of TPACK is extended from Shulman's (1986) idea of pedagogical content knowledge (PCK), which emphasizes that the paramount element of the teaching profession is the knowledge of how to make the subject matter more comprehensible to others. TPACK further promotes the position of the knowledge of teaching tools in the original framework of PCK to a higher level and stresses its interaction with the knowledge bases of subject matter and pedagogy as well (Mishra & Koehler, 2006). It asks teachers, regarding ICT adoption in teaching practice, to deliberate on how the subject matter might be represented by the application of ICT, how the teaching/learning process might be changed by the use of ICT, and, most importantly, how to make the subject matter more comprehensible to students with the aid of ICT. Although (as detailed in previous chapters) the constructs of TPACK are still open to debate, we consider that the main idea of TPACK is useful against the tendency of viewing ICT exclusively without weighing how it may serve teaching purposes.

Koehler, Mishra, and their colleagues advocated the use of the *learning-technology-by-design* approach (Koehler & Mishra, 2005; Koehler, Mishra, Hershey, & Peruski, 2004; Koehler, Mishra, & Yahya, 2007; Mishra & Koehler, 2006) to renovate the ICT courses in conventional teacher preparation programs. It is believed that the preservice teacher will start connecting his/her knowledge bases of subject matter, pedagogy, and ICT with each other when he/she is engaged in developing ICT solutions to pedagogical problems. We recognize that the nature of the development of ICT solutions to pedagogical problems should be intertwined with the transactional interplay between subject matter, pedagogy, and ICT. However, how to manage a course driven by the learning-technology-by-design approach and how to develop preservice teachers' sensitivity to the transactional interplay between subject matter, pedagogy, and ICT in such a course still remain big challenges for teacher educators. In this chapter, we introduce a course model that has been deployed in the science teacher education courses of National Taiwan Normal University (NTNU) for taking on the aforementioned challenges. The TPACK framework is adapted as an analytic tool to examine the growth in preservice science teachers' knowledge about technology integration in teaching.

6.2   The MAGDAIRE Model

The main challenges of deploying the learning-technology-by-design approach into an ICT course are how to ensure that preservice science teachers can generate ICT solutions to pedagogical problems within a given time period and how to keep them constantly revising their solutions to fulfill their pedagogical goals. In the NTNU courses, we engage preservice science teachers in collaboratively creating prototypes of ICT-integrated instructional materials for teaching scientific topics such as wind, solar and lunar eclipses, clouds, global warming, tides, tsunamis, typhoons, and rocks. Preservice science teachers are then asked to teach with these materials in the classroom. Numerous instructional design models have been proposed in the literature for helping teachers create instructional materials such as ADDIE (Dick & Carey, 1996) and ASSURE (Heinich, Molenda, Russell, & Smaldino, 2001). The general phases of instructional design include analysis, design/development, implementation, and evaluation (Reiser, 2001). These four phases are used as the systematic structure of our ICT courses to keep preservice science teachers on the track of creating ICT-integrated instructional materials.

Given that the learning-technology-by-design approach posits that the knowledge of ICT for teaching is fundamentally situated in the development of ICT solutions to pedagogical problems (Koehler et al., 2007), the cognitive apprenticeship should be the legitimate teaching strategy to support preservice teachers' inquiry into ICT. The cognitive apprenticeship reflects the learning perspective that knowledge is in part a product of the activity in which it is used. It refers to the reciprocal teaching between the expert and the novice during an authentic problem-solving activity just beyond what the novice can accomplish alone (Collins, 1988; Collins, Brown, & Newman, 1989). Within our ICT courses, Collins et al.'s (1989) teaching strategy of cognitive apprenticeships is revised and adopted as the pedagogical framework to scale up each phase of instructional design. The preliminary model of MAGDAIRE, which is abbreviated from modeled analysis, guided development, articulated implementation, and reflected evaluation, was proposed in Chien, Chang, Yeh, and Chang (2012). Based on previous studies (Chang, Chien, Chang, & Lin, 2012; Chien et al., 2012), we revised MAGDAIRE to better develop preservice science teachers' sensitivity to the interplay between subject matter, pedagogy, and ICT. Our 4-phase cyclic model is shown in Fig. 6.1. The mentoring team leading MAGDAIRE consists of several educational researchers, in-service teachers, and ICT experts.

6.2.1   Modeled Analysis

The main ICT tool emphasized for creating ICT-integrated instructional materials in the course is specified to the preservice teachers first. These students are explicitly informed that the main course tasks for them are, with the help of the mentoring team, to (a) create some prototypes of ICT-integrated instructional materials for their own future teaching practice and (b) teach with their prototypes before their

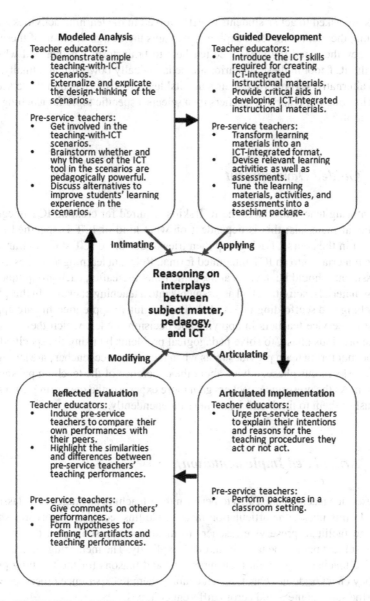

Fig. 6.1 MAGDAIRE model (Revised from Chien et al., 2012)

peers twice. They are then asked to form groups for collaboratively accomplishing the course tasks. The modeling facet of Collins' cognitive apprenticeship is applied in this phase to externalize the decision modes comprised of the interplay between subject matter, pedagogy, and ICT for the preservice teachers to imitate. To begin, the mentoring team demonstrates teaching-with-ICT scenarios of their own design for preservice teachers to experience. Within these scenarios, the preservice teachers are asked to act as students to get involved in the learning activities. At the same time, the mentoring team explicitly explains the design thinking of these scenarios; they then lead the preservice teachers to brainstorm whether and why the uses of the ICT tool in these scenarios are pedagogically powerful; and finally, they discuss alternatives to improve their students' learning experience in these scenarios. Each group of preservice teachers then selects a specific topic for teaching with the ICT tool.

6.2.2   Guided Development

The mentoring team introduces the ICT skills required for creating ICT-integrated instructional materials; this is dependent on what kind of ICT is specified as the major tool in the course. Each group then tries to use these ICT skills to transform learning materials into an ICT-integrated format. Relevant learning activities as well as assessments should be devised at the same time; thereafter, each group tunes the learning materials, activities, and assessments into a teaching package. In this phase, the coaching and scaffolding facets of Collins' cognitive apprenticeship are applied to assist preservice teachers in applying the decision modes, which they acquired from the previous phase, to solve pedagogical problems by using the specified ICT tool. The mentoring team provides hints when the preservice teachers are struggling with the tasks in this phase; advice about the coherence of the teaching packages is also given. As the preservice teachers gain more expertise, the mentoring team gradually pushes them to think and work more independently.

6.2.3   Articulated Implementation

The preservice teachers are asked to perform their teaching packages in a classroom setting. In this phase, the articulation facet of Collins' cognitive apprenticeship is applied to facilitate preservice teachers to make their reasoning on the interplay between subject matter, pedagogy, and ICT explicitly. The mentoring team asks the preservice teachers to explain their intentions and reasons for the teaching procedures they enact or do not enact, which enables them to experience others' perspectives in the same context and across different contexts.

6.2.4   Reflected Evaluation

The preservice teachers are asked to compare their own performance with those of their peers and then to comment on others' performances. The mentoring team highlights the similarities and differences between the teaching performances.

In this phase, the reflection facet of Collins' cognitive apprenticeship is applied to facilitate preservice teachers to modify their reasoning modes of the interplay between subject matter, pedagogy, and ICT. By involving peer assessment, each preservice teacher becomes a case for others to reconsider what elements might be critical to successful and unsuccessful teaching. Moreover, the comments from peers can function as a replay for them to reanalyze their own performances. It helps preservice teachers to form hypotheses for refining their ICT artifacts and teaching performance.

It should be noted that the reflected evaluation phase takes place as the formative assessment and triggers the next cycle of MAGDAIRE. As shown in Fig. 6.1, the second round of MAGDAIRE starts at the guided development phase. The leading position in the cognitive apprenticeship should gradually shift from the mentoring team to the preservice teachers as they become more skilled team members. The main task of the mentoring team should put more emphasis on encouraging preservice teachers to iteratively test the hypotheses that they form in the reflected evaluation phase.

6.3   Studies to Evaluate the Effectiveness of MAGDAIRE

MAGDAIRE has been deployed in NTNU science teacher education courses since 2010. Each course lasts for 18 weeks. In the courses conducted in 2010 and 2011, Adobe® Flash® was chosen as the main tool to develop ICT-integrated instructional materials. The reasons behind this choice were as follows: (a) Flash can compile static images into dynamic animations through automatic procedures. It may reduce the threshold of multimedia development for preservice teachers; (b) Flash can add functions into animations to make them interactive. It may help preservice teachers to build tools to assist students in visualizing, sharing, and testing ideas; (c) Flash enables animations to record, retrieve, and exchange user information on the Internet. It may benefit preservice teachers in tracking students' learning progress; (d) Flash-made content is accessible to various computer systems and mobile devices (Adobe, 2014). The Flash-integrated instructional materials made by preservice science teachers should be usable for their future teaching. Several studies have been conducted along with the courses driven by MAGDAIRE. In the following paragraphs, we summarize the key findings of two of these studies (Chang et al., 2012; Chien et al., 2012).

Within our courses, preservice science teachers had to go through the MAGDAIRE model twice in one semester. They were asked to write down, periodically, their ideas about how to revise their ICT-integrated instructional materials and the reasons behind their decision as well. As shown in Table 6.2, Koehler et al.'s (2007) framework of TPACK was revised to analyze the preservice science teachers' reasoning patterns regarding the interplays between subject matter, pedagogy, and the use of Flash. Two main categories emerged to represent preservice teachers' modes of reasoning about revision in Flash-integrated instructional materials, including

Table 6.2 Coding protocol of preservice teachers' reasoning patterns

Main category	Subcategory (code)	Exemplar
Isolated Content (C) Pedagogy (P) Flash (F)	Content (C): reasoning on the actual science subject matter that is to be taught such as clarifying the facts, concepts, and theories of the chosen subject matter	We have to clarify the explanation for the greenhouse effect. When solar radiation passes through the earth's atmosphere, it warms the planetary surface. The greenhouse gases in the atmosphere absorb the infrared thermal radiation emitted from the planetary surface. Furthermore, the greenhouse effect already exists before the occurrence of the so-called global warming
	Pedagogy (P): reasoning on the processes and methods of teaching and learning; furthermore, how it encompasses overall educational purposes, values, and aims such as arranging students' learning steps	We should administer a test to students by the end of the course. It can help us to understand students' learning progress and offer us information to give students appropriate feedback
	Flash (F): reasoning on the use of Flash but not specifically related to the chosen subject matter or teaching strategies such as the operation of one particular function of Flash	The animation may suddenly break off while playing. Maybe we should use the frame-by-frame approach to compile the animation
Joint Content pedagogy (CP) Flash content (FC) Flash pedagogy (FP) Flash content pedagogy (FCP)	Content pedagogy (CP): reasoning on how particular aspects of a science subject matter are organized, adapted, and represented for instruction such as specifying one teaching strategy to complement one particular concept of the chosen subject matter but not related to the use of Flash	By comparing with other subjects in the domain of Earth Science, the learning unit of rocks puts more emphasis on students memorizing the facts. In addition to introducing the characteristics of varied kinds of rocks and the Mohs hardness scale, we should encourage students to compare the differences in the hardness, crystal system, crystal class, and streak between minerals to enhance their learning motivation
	Flash content (FC): reasoning on how the chosen science subject matter might be shaped by the application of Flash such as leveraging one particular function of Flash to present the chosen subject matter	Pictures of clouds can be embedded with more detailed information by utilizing ActionScript. We attempt to revise the scripts to make the clouds' characteristics such as cloud classification and cloud height appear when the cursor is moved onto the cloud pictures

(continued)

Table 6.2 (continued)

Main category	Subcategory (code)	Exemplar
	Flash pedagogy (FP): reasoning on how teaching as well as learning might be changed by the use of Flash such as leveraging one particular function of Flash to support one particular teaching strategy	The online testing system should be added, with ActionScript, to count students' scores. If a student enters wrong answers to a question too many times, the system will automatically force him/her to view the animation that explains the concept of the question
	Flash content pedagogy (FCP): reasoning on how the chosen science subject matter might be shaped by the application of Flash and the impact of such applications on teaching methods such as leveraging one particular function of Flash to present the chosen subject matter and support one particular teaching strategy	We plan to add interactive functions into the animation depicting sea waves to allow students to manipulate the variations in water depth along the coastline. Then, students can test their hypotheses of the relation between water depth and wave speed and then discuss the data they obtain from the animation with peers. It will facilitate students in exploring the relationship between water depth and wave speed. It also helps to explain the differences between deep and shallow waves

Adapted from Chang et al. (2012)

(a) isolated modes, indicating that content, pedagogy, or Flash were addressed in isolation, and (b) joint modes, indicating that content, pedagogy, and Flash were treated as intertwined elements. It was found that, as shown in Fig. 6.2, preservice teachers' reasoning patterns showed profound changes toward a more joint mode across time; they became aware that teaching with Flash should be a work blending subject matter, pedagogy, and Flash. As shown in Fig. 6.3, when compared to the first round of MAGDAIRE, the percentage of the summation of joint patterns at the second round of MAGDAIRE significantly increased from 50 to 80 % ($p < .05$).

Semistructured interviews were conducted to retrospectively infer possible mechanisms that facilitated the changes in preservice science teachers' reasoning patterns regarding the interplays between subject matter, pedagogy, and the use of Flash. The main themes of the interviews that relate to this chapter were (a) the contradictions in the preservice teachers' instructional planning processes within the context of MAGDAIRE and the solution to the said problems (e.g., what difficulty did you encounter in this course and how did you resolve it?) and (b) the preservice teachers' perceptions of the usefulness of the ICT-integrated materials they produced within the context of MAGDAIRE (e.g., will you implement your ICT-integrated materials in your future teaching practice? Why or why not?). An inductive analysis was conducted on the interview data. The documents of the preservice teachers' practice, including weekly coursework, videotaping for group presentations, discussion in an online forum, and comments on peers' work, were used to examine and refine the interpretation of the interview data.

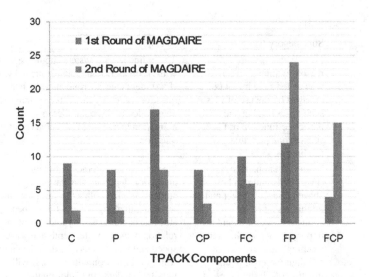

Fig. 6.2 Distribution of preservice teachers' reasoning patterns over phases (Data source of this figure: Chang et al., 2012). *C* content, *P* pedagogy, *F* flash, *CP* content pedagogy, *FC* flash content, *FP* flash pedagogy, *FCP* flash content pedagogy

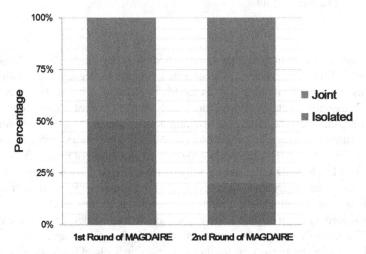

Fig. 6.3 Percentages of the isolated and joint patterns over phases (Data source of this figure: Chang et al., 2012)

The case of John's group from Chien et al. (2012) illustrates the changes in these preservice science teachers' reasoning and its relation to their practice. John's group chose the topic "typhoon" from the Taiwanese high school Earth Science textbook as the subject matter to be taught. Within the modeled analysis phase, they decomposed the teaching-with-ICT scenarios, which were demonstrated by the mentoring team, to justify why and how Flash should be used in teaching. John's group thought that Flash would be powerful to compile the textbook's static diagrams with explanatory text into animations. They believed that animations would attract students' attention and make the topic more comprehensible. Such simple means–end connections between the use of Flash and possible consequences became their major principle of ICT development; they tried to transform the content of the textbook into animations whenever possible in the guided development phase. However, John's group felt pretty depressed in the articulated implementation phase because the reactions of peers were far from what they expected. John's group had the impression that other preservice science teachers were indeed attracted by the transition effects of animations during the microteaching session, but they did not pay much attention to the scientific explanations embedded in the animations. Furthermore, John's group recognized that, while using Flash to compile animations was not difficult, it was a time-consuming process. They were afraid that, within the limited course time, their final ICT-integrated instructional materials would become fancy but superficial if they just kept focusing on transforming the entire textbook material into animations.

In the reflected evaluation phase, John's group came up with another approach to keep students' attention: constantly posing questions to students. This decision fundamentally changed the structure of their ICT-integrated instructional materials. They started reconsidering what concepts could be intertwined to form a series of interrelated questions and then decided to put more emphasis on the complex interaction between typhoon, topography, wind direction, and rainfall. The question *Why can a typhoon bring about various rainfalls over different locations?* was set as the main question driving the whole teaching procedure. In the second round of the guided development phase, John's group shifted their efforts to create an animation that depicted various typhoon pathways on a Taiwan map. As shown in Fig. 6.4, some interactive functions were added to the animation, enabling students to manipulate typhoon pathways. The data about rainfall and wind direction of typhoons Matsa, Haytang, and Dujuan (which struck Taiwan in 2003 and 2005) in different areas were also embedded in the animation. It was found that John's group's use of Flash became more content specific. In the second round of the articulated implementation phase, the main theme of John's group was to engage students in forming hypotheses about relations between rainfall, wind direction, and landforms. The role of animations was repositioned as the tool for students to generate answers rather than reading materials only. The interactive models of typhoons Matsa, Haytang, and Dujuan embedded in the animation were used to test students' hypotheses. The form of classroom activities became more interactive; it shifted from lecturing with Flash to interacting with Flash. As for the perspective of motivation empowerment, the value toward the use of Flash changed from "just for fun" to helping students to think harder.

Through reviewing peer performance, John's group noticed that the one-way linear connection between the three typhoon models would have difficulty in accommodating each student's learning pace as well as the teacher's teaching pace. They then increased referential links and nodes among each section of their ICT-integrated instructional materials to form various operating pathways. The method of

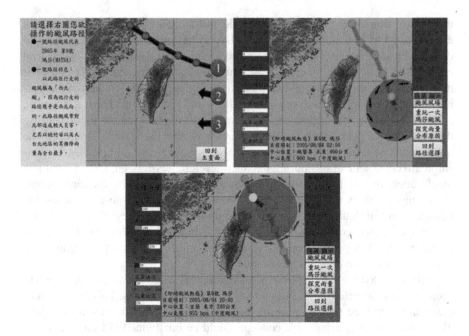

Fig. 6.4 Screenshots of John's group's ICT-integrated instructional materials

information presentation transformed from linear to nested formats. However, they were still concerned about whether students could be well prepared for future Earth Science tests from their teaching. After the second round of the reflected evaluation phase, John's group tried to add more systematic guidance and drill-and-practice exercises into the animations to help students prepare for tests. Their perceptions of teaching with ICT seemed to be in the transition between instructivist orientation and constructivist orientation. In sum, John's group reexamined the connections between the use of Flash and their practice from the views of subject matter selection, motivation empowerment, information presentation, activity design, and pedagogy transition. This case suggested that MAGDAIRE could evolve preservice science teachers' TPACK. Moreover, it stimulated these preservice science teachers to modify their reasoning modes and, consequently, revise their practice. Other cases delineating the changes in preservice science teachers' reasoning and practice within MAGDARE can be found in Chien et al. (2012).

6.4 Concluding Remarks

In this chapter, we introduced the MAGDAIRE model for renovating ICT courses in conventional teacher preparation programs. The TPACK framework was adapted as an analytic tool to examine the growth in preservice science teachers' knowledge about technology integration in teaching. The key findings indicated that MAGDAIRE facilitated preservice science teachers to reexamine the connections between the use of ICT and their teaching practice. Moreover, MAGDAIRE significantly enhanced preservice teachers' sensitivity to the transactional interplay between subject matter, pedagogy, and ICT. It is worthy to note that several of the preservice science teachers who participated in MAGDAIRE voluntarily entered the 2012 ICT-integrated microteaching competition that was held by the NTNU Office of Teacher Education and Career Service. Their teaching performances were appreciated by in-service teachers as well as teacher educators and won first and second prizes in the competition. This encouraging news makes us confident that MAGDAIRE has a positive and practical impact on preservice science teachers.

References

Adobe. (2014). *Adobe Flash runtimes statistics.* Retrieved March 25, 2015, from http://www.adobe.com/tw/products/flashruntimes/statistics.html

Angeli, C. (2005). Transforming a teacher education method course through technology: Effects on preservice teachers' technology competency. *Computers & Education, 45*(4), 383–398.

Chang, C.-Y., Chien, Y.-T., Chang, Y.-H., & Lin, C.-Y. (2012). MAGDAIRE: A model to foster pre-service teachers' ability in integrating ICT and teaching in Taiwan. *Australasian Journal of Educational Technology, 28*(6), 983–999.

Chien, Y.-T., Chang, C.-Y., Yeh, T. K., & Chang, K.-E. (2012). Engaging pre-service science teachers to act as active designers of technology integration: A MAGDAIRE framework. *Teaching and Teacher Education, 28*(4), 578–588.

Collins, A. (1988). *Cognitive apprenticeship and instructional technology.* (Report No. BBN-R-6899). Cambridge, MA: BBN Laboratories. Retrieved from ERIC Database. (ED 331 465).

Collins, A., Brown, J. S., & Newman, S. E. (1989). Cognitive apprenticeship: Teaching the crafts of reading, writing, and mathematics. In L. B. Resnick (Ed.), *Knowing, learning, and instruction: Essays in honor of Robert Glaser* (pp. 453–494). Hillsdale, NJ: Lawrence Erlbaum.

Dick, W., & Carey, L. (1996). *The systematic design of instruction* (4th ed.). New York: HarperCollins.

Govender, D., & Govender, I. (2009). The relationship between information and communications technology (ICT) integration and teachers' self-efficacy beliefs about ICT. *Education as Change, 13*(1), 153–165.

Heinich, R., Molenda, M., Russell, J. D., & Smaldino, S. E. (2001). *Instructional media and technologies for learning* (7th ed.). Englewood Cliffs, NJ: Prentice Hall.

International Society of Technology in Education. (2008). *National educational technology standards for teachers.* Eugene, OR: Author.

Jang, S.-J., & Chen, K.-C. (2010). From PCK to TPACK: Developing a transformative model for pre-service science teachers. *Journal of Science Education and Technology, 19*(6), 553–564.

Koehler, M. J., & Mishra, P. (2005). Teachers learning technology by design. *Journal of Computing in Teacher Education, 21*(3), 94–102.

Koehler, M. J., Mishra, P., Hershey, K., & Peruski, L. (2004). With a little help from your students: A new model for faculty development and online course design. *Journal of Technology and Teacher Education, 12*(1), 25–55.

Koehler, M. J., Mishra, P., & Yahya, K. (2007). Tracing the development of teacher knowledge in a design seminar: Integrating content, pedagogy and technology. *Computers & Education, 49*(3), 740–762.

Mahdizadeh, H., Biemans, H., & Mulder, M. (2008). Determining factors of the use of e-learning environments by university teachers. *Computers & Education, 51*(1), 142–154.

Ministry of Education. (2001). Information education report [in Chinese]. Taipei, Taiwan: Author. Retrieved from http://content.edu.tw/primary/info_edu/tp_tt/content/nerc-1/law/teacher_point. htm

Ministry of Education. (2008). *White paper on information technology education for elementary and junior high schools 2008–2011*. Taipei, Taiwan: Author.

Mishra, P., & Koehler, M. J. (2006). Technological pedagogical content knowledge: A framework for teacher knowledge. *Teachers College Record, 108*(6), 1017–1054.

Niess, M. L. (2005). Preparing teachers to teach science and mathematics with technology: Developing a technology pedagogical content knowledge. *Teaching and Teacher Education, 21*(5), 509–553.

Organization for Economic Cooperation and Development. (2008, May). *New millennium learners: Initial findings on the effects of digital technologies on school-age learners*. Paper presented at the OECD/CERI International Conference on Learning in the 21st Century: Research, Innovation and Policy, Paris, France. Retrieved from http://www.oecd.org/site/educeri21st/40554230.pdf

Organization for Economic Cooperation and Development. (2010). *Inspired by technology, driven by pedagogy: A systemic approach to technology-based school innovations*. Paris: Author. doi:10.1787/9789264094437-en

Pelgrum, W. J., & Doornekamp, B. G. (2009). *Indicators on ICT in primary and secondary education* (Report No. EACEA-2007-3278/001-001). Retrieved from Education, Audiovisual & Culture Executive Agency website: http://eacea.ec.europa.eu/llp/studies/documents/study_on_indicators_on_ict_education/final_report_eacea_2007_17.pdf

Reiser, R. A. (2001). A history of instructional design and technology – Part II: A history of instructional design. *Educational Technology Research and Development, 49*(2), 57–67.

Shulman, L. S. (1986). Those who understand: Knowledge growth in teaching. *Educational Researcher, 15*(2), 4–31.

Sørebø, Ø., Halvari, H., Gulli, V. F., & Kristiansen, R. (2009). The role of self-determination theory in explaining teachers' motivation to continue to use e-learning technology. *Computers & Education, 53*(4), 1177–1187.

Thompson, A. D., & Mishra, P. (2007–2008). Breaking news: TPCK becomes TPACK! [Editorial]. *Journal of Computing in Teacher Education, 24*(2), 38 & 64.

Wilson, E. K. (2003). Preservice secondary social studies teachers and technology integration: What do they think and do in their field experiences. *Journal of Computing in Teacher Education, 20*(1), 29–39.

Chapter 7
Examining Teachers' TPACK in Using e-Learning Resources in Primary Science Lessons

Winnie Wing-Mui So, Apple Wai-Ping Fok, Michael Wai-Fung Liu, and Fiona Ngai-Ying Ching

The advocates of technology in education have dramatically stirred up the life of teachers, requiring substantial changes to their practices and processes of teaching and learning. Yet, there are tendencies to merely introduce technology to teaching and learning without much understanding of the knowledge required for teachers to use the technology effectively and efficiently. This study refers to the technological pedagogical content knowledge (TPACK) framework to better understand the phenomenon of teachers' integration of content knowledge, pedagogy, and technology in their teaching. In this study, e-learning resources of four science topics in Key Stage 2 of the Hong Kong primary curriculum have been designed and developed based on the resource-based e-learning environments (RBeLEs). A total of 19 teachers from six primary schools were invited to use these e-learning resources in their classrooms. Analysis of the teachers' use of the e-learning resources can help to provide tangible understanding of how technology supports teaching and learning. The data collected included students' pre/post lesson tests, lesson observations of teachers' use of the e-learning resources, and teachers' interview responses that provided useful information to enhance our understanding of how e-learning resources are used in primary classrooms.

W.W.-M. So (✉) • F.N.-Y. Ching
The Hong Kong Institute of Education, Hong Kong, Hong Kong
e-mail: wiso@ied.edu.hk

A.W.-P. Fok
City University of Hong Kong, Hong Kong, Hong Kong

M.W.-F. Liu
Ying Wa Primary School, Hong Kong, Hong Kong

© Springer Science+Business Media Singapore 2015   105
Y.-S. Hsu (ed.), *Development of Science Teachers' TPACK*,
DOI 10.1007/978-981-287-441-2_7

7.1 Introduction

In recent years, the advocacy of the integration of technology into teaching in primary and secondary education has greatly influenced teachers, urging them to make substantial changes to their practices and processes of learning and teaching. The 2011–2012 Policy Address by the Chief Executive of Hong Kong clearly expressed the Government's commitment to developing e-learning resources:

> Developing Electronic Textbooks
> The use of e-learning resources has become a major trend in education. Apart from providing students with an interactive mode of learning, electronic textbooks and learning resources allow more flexibility in textbook compilation, lower production costs, reduce wastage and help achieve reasonable pricing. This is a desirable alternative to printed textbooks, which is currently the only option available on the market.
> The Government is committed to developing e-learning resources. An EDB [Education Bureau] task force set up in mid-2011 will review teaching and learning materials and explore ways to better utilize the advantages of e-learning and improve the provision of textbooks. (2011, para. 120 & 121)

7.1.1 Technological Pedagogical Content Knowledge (TPACK)

Using electronic textbooks, providing e-learning resources, or simply adding technology to old ways of teaching is insufficient to enhance teaching and learning. The key is teachers' knowledge of integrating technology, pedagogy, and content. Building upon Shulman's (1986) pedagogical content knowledge (PCK), the technological pedagogical content knowledge (TPACK) framework is developed for a better understanding of the kinds of knowledge teachers need to teach effectively with technology (Koehler & Mishra, 2005; Mishra & Koehler, 2006). The framework focuses on the "dynamic, transactional relationship between content, pedagogy, and technology" (Koehler, Mishra, & Yahya, 2007, p. 741), emphasizing that effective integration for teaching subject matter requires knowledge not just of content, technology, and pedagogy but also of their relationship to each other (Fig. 7.1). Since teaching is a context-bound activity, it is also necessary to take learning contexts into consideration in the discussion of TPACK (Koehler, Mishra, Akcaoglu, & Rosenberg, 2013).

7.1.2 Resource-Based e-Learning Environments (RBeLEs)

The resource-based e-learning environments (RBeLEs) framework was developed by making reference to the two discussions about learning environments—resource-based learning environments (Hill & Hannafin, 2001) and sciences-based learning environments (Blumenfeld, Kempler, & Krajcik, 2006)—and by drawing from the

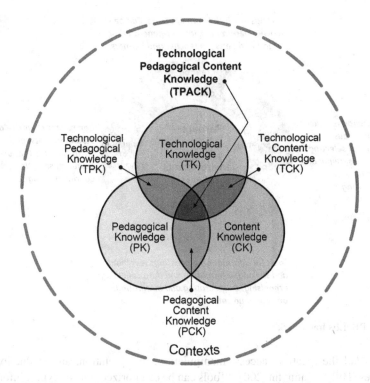

Fig. 7.1 TPACK framework (Reproduced with permission from http://tpack.org)

findings of a study that aimed to understand how teachers made use of online resources to design learning environments (So, 2012). RBeLEs provide a set of systematic and detailed guidelines for teachers to construct e-learning environments with Internet resources for student learning. It consists of four features: creation of contexts, selection of resources, use of tools, and adoption of scaffolds (Fig. 7.2). Each feature and its connection to learners' motivation and cognitive engagement are discussed below.

Creation of contexts refers to the creation of the settings in which learning occurs. Contexts can be determined by the teacher, generated by the learners, or negotiated between the teacher and learners. Contexts need to be authentic and closely related to learners' everyday life to be motivating. If the contexts are generated by the learners or negotiated, they could increase interest (intrinsic value) in learning and a sense of autonomy.

Resources contain the core information presented in resource-based learning environments (Hill & Hannafin, 2001). They can be dynamic or static. Dynamic resources undergo frequent, sometimes continual, change. Static resources, in contrast, have relatively stable contents that may quickly become obsolete or inaccurate. Resources need to be relevant and of an appropriate level for the learners to sustain interest and promote cognitive engagement. Allowing learners to have considerable control over the computers and resources also increases their perceived competence and sense of autonomy.

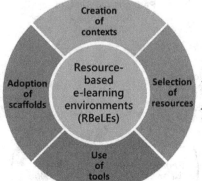

Settings in which learning occurs. Can be
teacher-determined, learner-generated or
negotiated between teacher and learners.

Can be categorised into
asking and discussion;
searching and selecting;
doing and observing; and
summarising and
conceptualising.

Can be static (with relative
stable contents) or dynamic
(with contents that undergo
frequent, sometimes
continual change).

Tools for information processing; searching
and seeking; information and data collecting;
organising; collaborating and integrating;
and communicating.

Fig. 7.2 RBeLEs framework

Tools aid the location, access, manipulation, interpretation, and evaluation of resources (Hill & Hannafin, 2001). Tools can be categorized into six types: information processing, searching and seeking, information and data collection, organizing, collaborating and integrating, and communicating. The provision of tools to learners promotes their perceived competence to succeed in learning tasks and, therefore, creates engaging experiences for them. Some of the tools can also increase learning interest, foster relatedness, and enhance the learners' sense of autonomy.

Scaffolds provided by teachers are necessary and important in sustaining learners' interest, helping them to become more willing to approach challenges, construct understanding, and have a positive inquiry experience with their peers (Blumenfeld et al., 2006). There can be four types of scaffolds: asking and discussing, searching and selecting, doing and observing, and summarizing and conceptualizing. Scaffolding provided by teachers can help learners become more willing to approach challenges by increasing their interest in the topic and their perceived competence. With an appropriate degree of scaffolding, learners can experience engagement when performing a task accompanied by an increased sense of relatedness and autonomy.

7.2 Methodology

The main purpose of this study is to better understand how primary teachers integrate CK, pedagogy, and technology (including tools and resources) in their teaching through the use of multimedia resources on an online learning platform.

7.2.1 Participants and Settings

Altogether, 19 teachers teaching general studies (GS) and their Grades 4 and 5 students ($N=540$) took part in this study. General studies (GS) is a core subject of the primary school curriculum in Hong Kong. This interdisciplinary subject provides primary school children with opportunities to integrate knowledge, skills, values, and attitudes across the key learning areas of personal development, social and humanities education, science education, and technology education (Curriculum Development Council, 2011). In accordance with the curriculum reform and development, GS focuses on student inquiry and developing students' ability and skills for learning to learn and has an emphasis on the use of diverse resources. Hence, this subject provides a great opportunity for e-learning.

7.2.2 e-Learning Resources on the Web-Based Learning Platform

The research team, consisting of curriculum experts and teacher educators in GS, created a context for RBeLEs with a selection of appropriate resources available on the web as well as suggestions for the use of tools and the adoption of scaffolds. In this study, the teaching and learning materials were arranged and delivered on an online teaching and learning platform. In the past, e-learning systems focused on the transfer of knowledge and information; currently, they aim to support collaborative learning and constructivism and to promote students' locus of control. Based on the content designed by the research team, the technical team created a series of flash animations within which the e-learning resources were embedded.

Web-based interactive instructional materials that could be used for inquiry learning and practicing purposes including lesson plans, tutorials, audiovisual aids, and games were designed according to the set of learning objectives that had been determined. These materials would not only be used for presentation, learner practice, and communicative interaction but, more importantly, as a source of stimulation and inspiration for classroom activities. Taking advantage of the web-based learning environment, interactions between students and students, students and teachers, and students and materials could be broadened and deepened. Knowledge construction is a complex process that results from learning that is derived from the mediating interaction among the materials/activities. This demonstrates a systematic approach that is useful in explaining the role of materials in the overall curriculum and building/strengthening teaching confidence in using interactive tools and materials.

The online learning platform offers a variety of e-learning resources and tools to support teaching and learning, including online discussion forum, online voting, assessment engine, animation/online videos, reading materials, and interactive/simulation games. Examples for each of these resources are shown in Figs. 7.3, 7.4, 7.5a, 7.5b, 7.5c, 7.6a, 7.6b, 7.7a, 7.7b, 7.7c, and 7.8 with screen captures.

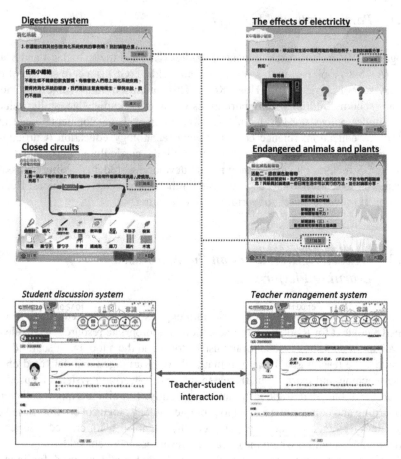

Fig. 7.3 Online discussion forum

Online discussion forum (Fig. 7.3): Students exchange views on a certain topic on the online discussion forum with other students and the teacher. The teacher can monitor students' postings and provide feedback.

Online voting (Fig. 7.4): Students cast their votes via the online voting system. The teacher can display the results immediately and discuss them with the students.

Assessment engine: The online learning platform provides three types of assessment:

- Making predictions prior to observation (Fig. 7.5a)—Students make predictions on the online learning platform, then observe the results, and finally compare the results they observe with what they predicted.
- Selection/oral response (Fig. 7.5b)—Students select answers on the online learning platform by clicking, drag and drop, or matching. The teacher can also ask

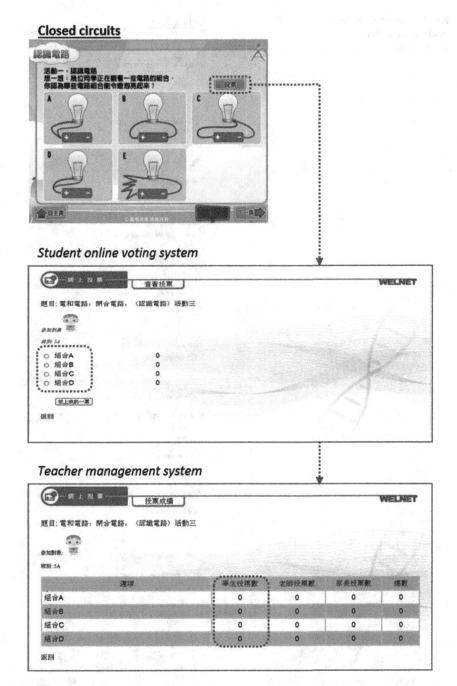

Fig. 7.4 Online voting

Fig. 7.5a Assessment engine
(making predictions prior to
observation)

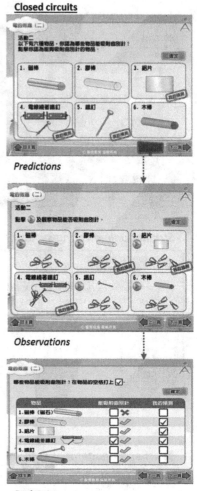

Fig. 7.5b Assessment engine (selection/oral response)

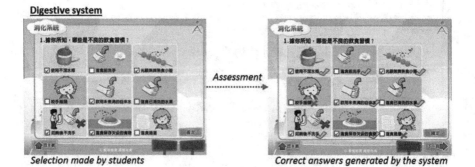

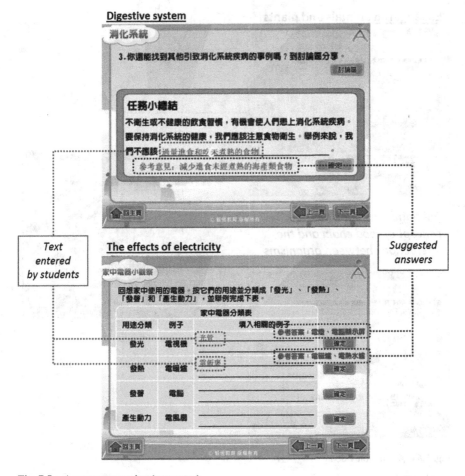

Fig. 7.5c Assessment engine (text entry)

students to give their answers orally. Assessment can be made by using the correct answers generated by the system.

- Text entry (Fig. 7.5c)—Students enter text into the online learning platform to answer questions or express their views.

Animation/online videos: Students view either the animation (Fig. 7.6a) or an online video (Fig. 7.6b) that introduces a concept or explains a phenomenon.

Reading materials: The online learning platform provides three types of reading materials including websites (Fig. 7.7a), news information/extended materials (Fig. 7.7b), and case diagrams (Fig. 7.7c). Students obtain information from the reading materials to increase their understanding of a certain topic.

Interactive/simulation games (Fig. 7.5a): Students carry out virtual experiments through playing simulation games.

Endangered animals and plants

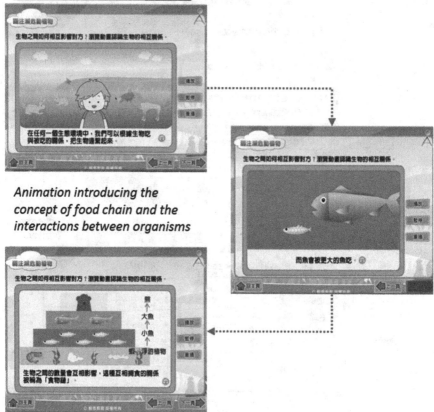

*Animation introducing the
concept of food chain and the
interactions between organisms*

Fig. 7.6a Animation video

Digestive system

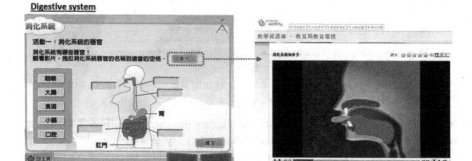

Online video introducing the digestive system

Fig. 7.6b Online video

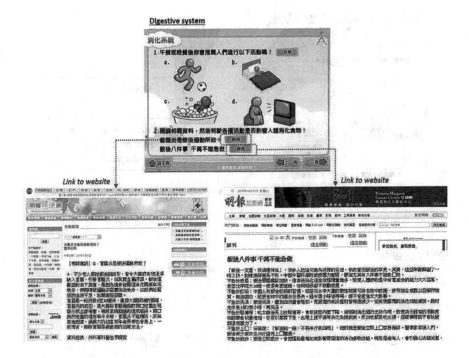

Fig. 7.7a Reading materials (websites)

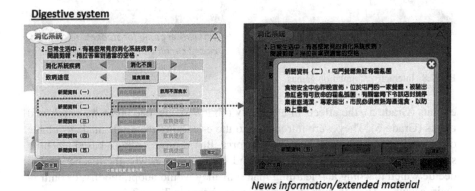

News information/extended material

Fig. 7.7b Reading materials (news information/extended materials)

7.2.3 Teaching Preparation

All 19 teachers participated in a workshop that introduced them to the concept of TPACK as well as RBeLEs. They then had lesson planning sessions with the researchers on specific grade-level topics: the digestive system (Grade 4), closed

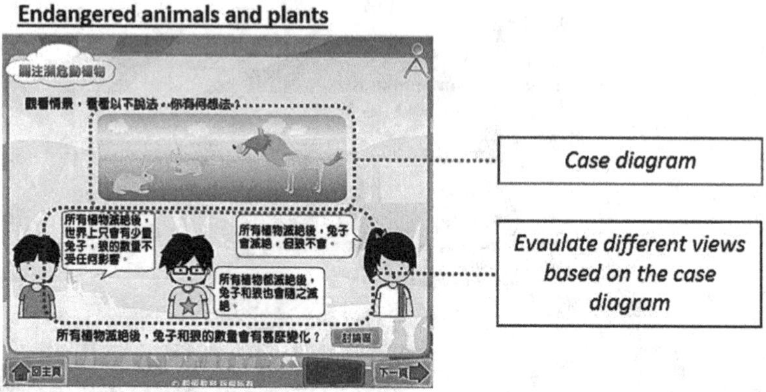

Endangered animals and plants

Fig. 7.7c Reading materials (case diagrams)

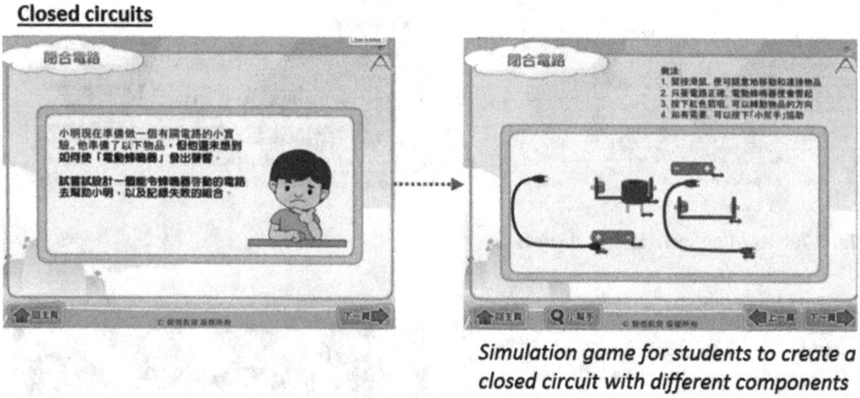

Closed circuits

Simulation game for students to create a closed circuit with different components

Fig. 7.8 Interactive/simulation games

circuits (Grade 5), the effects of electricity (Grade 5), and endangered animals and plants (Grade 6). They were introduced to the e-learning materials made available on the online teaching and learning platform. Based on the school-based teaching schedule, the students' characteristics, and the teaching needs, the teachers could decide which teaching and learning materials they would adopt, modify, or omit in their teaching. The contents of each topic are shown in Table 7.1.

7.2.4   Data Collection and Analysis

7.2.4.1   Lesson Observations

All the lessons were video-taped by the teachers themselves and later viewed by the researchers to record the percentage of use of different multimedia resources and how each resource was used. For ease of comparison, the percentage of use of each

Table 7.1 Grades, topics, and activities involved in the study

Grade	Topic	Activity
4	The digestive system	Digestive organs
		Work of the digestive system
		Functions of the digestive system
		Importance of chewing
		Healthy diet
		Protecting the digestive system
5	Closed circuits	Dry cells
		Electrical circuits
		Conductors and nonconductors of electricity
6	The effects of electricity	Observing small electrical appliances at home
		Light, sound, and magnetic effects of electricity
7	Endangered animals and plants	Relationships between living organisms
		Interactions between living organisms
		Searching for endangered animals and plants
		Causes of endangerment
		Saving endangered animals and plants
		A dilemma

resource was obtained by dividing the number of times the resource was used as recorded during the lesson observations by the number of times the resource was available for use on the online platform. For example, the online discussion forum appeared 18 times in the online platform across the four topics learned by the six classes but was used only thrice. So, dividing 3 by 18, the percentage of use of the online discussion forum by the teachers was 17 %.

7.2.4.2 Teacher Interviews

To gain a deeper understanding of how the teachers applied their understanding of TPACK in their teaching, an interview protocol consisting of 12 questions was developed. The 19 teachers were interviewed face-to-face individually after the lessons. All interview sessions were audio-recorded and later transcribed for analysis. The questions focused on two main areas regarding their use of technology in teaching: technology-supported pedagogical knowledge (PK) and technology-related classroom management knowledge.

7.2.4.3 Student Pre- and Postlesson Tests

To help teachers evaluate the effects of student learning with multimedia resources on the online learning platform, the researchers worked collaboratively with teachers to design pre- and postlesson tests for each topic. These tests consisted of the same set of questions. Students completed the tests at around 1 week before and after the lesson. Paired samples *t*-tests were used to compare changes in students' understanding of the topic before and after the lesson.

7.3 Findings and Discussion

7.3.1 Lessons with e-Learning Resources: Preparation

During the postlesson interviews, the teachers were asked to state three factors they had considered when designing the teaching and learning activities that involved the use of technology. These factors were found to be related to the teachers' PCK and the knowledge domains surrounding it. Most teachers stated that how well they understood the e-learning materials and how much confidence they had in the topic as factors they would consider. These factors were related to the teachers' CK, PK, and PCK. As an extension of PCK, TPACK is embedded in an educational context (Doering, Veletsianos, Scharber, & Miller, 2009) and influenced by a number of knowledge domains—learners, schools, subject matter, curriculum, and pedagogy (Niess, 2001, 2011).

In this study, factors related to these knowledge domains were reorganized into three dimensions. For the student dimension, they were concerned about students' information communication technology (ICT) literacy and the number of students in each class. Apart from considerations about the teachers themselves and their students, the teachers would also take the school dimension into account when designing teaching and learning activities involving the use of technology. The teachers were most worried about whether implementing such activities would affect the teaching schedule and, consequently, students' examination results. Another consideration regarding the school dimension was the adequacy of different technology equipment: most of the schools have only one computer room and a few dozen tablets or laptops to be shared by all the students. The teachers also emphasized the importance of whether the e-learning materials could match the content of the textbooks or school-based curriculum and whether they could enhance students' learning efficacy and motivation.

However, the teachers hardly mentioned any factors related to their own technological knowledge (TK), technological content knowledge (TCK), or technological pedagogical knowledge (TPK). This might be due to the fact that the e-learning materials were already determined and the teachers' responsibilities were to design lessons with the different materials available on the online platform.

Since students' ICT literacy was one of the factors that the teachers would consider when designing e-learning activities, they were asked how they assessed students' ICT literacy. Most teachers said they relied on observations and previous teaching experiences; others said they consulted the students' ICT teacher about their ICT literacy so as to assess the appropriateness of the e-learning activities. The teachers observed that, although most of the primary school students have experience using online social networking sites (e.g., Facebook) and playing online games, they are not very interested in online learning platforms. Moreover, although students have basic word processing skills, their Chinese typing skills are in general very limited; thus, students with lower ability would need more instruction and time to be able to navigate effectively around the online learning platform.

Other than collecting information about students' ICT literacy, most of the teachers made special preparations before using technology in their lessons. They tried to familiarize themselves with the e-learning materials, make more detailed plans for the flow of teaching, and make contingency plans in case of technology failure. The teachers also checked the hardware and e-learning materials before the lesson. It was observed that these teachers' preparation work was more focused on the technical and teaching content levels rather than on the pedagogical level.

7.3.2   Lessons with e-Learning Resources: Implementation

7.3.2.1   Types and Percentage of Use of the e-Learning Resources

The data from the lesson observations show that some e-learning resources were more popular among some teachers while others were less frequently used. By making reference to the RBeLEs framework, e-learning resources available on the online learning platform were placed in two categories: functional tools and multimedia resources.

Functional tools include the online discussion forum, online voting, and the assessment engine (i.e., making predictions prior to observation, selection/oral response, text entry). The online discussion forum and online voting are communication tools that enable the sharing and exchange of ideas. The assessment engine is an information processing tool that provides cognitive support for information management. Multimedia resources include animation/online videos, reading materials (i.e., websites, news information/extended materials, case diagrams), and interactive/simulation games. The animation/online videos and reading materials selected for the topics of this study are mostly static; that is, they have stable contents. The interactive/simulation game is dynamic as it changes when the user interacts with it. The percentage of use of each resource by the 19 teachers is listed in Table 7.2.

Of the functional tools, the online discussion forum was used the least (17 %) by the teachers; it was also the least frequently used among all types of e-learning resources. Data from the teacher interviews revealed that these teachers believed that most primary school children have very limited Chinese typing skills, which would hinder their effective participation in online discussion forums. Therefore, the teachers would rather use worksheets, teacher-led discussion, student group discussion, and group presentations.

The percentage of online voting usage was 46 %. The teachers who chose to use this resource said that online voting was easy for primary school students to handle and that using it would increase participation. However, the teachers who did not

Table 7.2 Teachers' use of each type of e-Learning resource

e-Learning resource	Percentage of use
Functional tools	
Online discussion forum	16.7
Online voting	45.5
Assessment engine	
Making predictions prior to observation	33.3
Selection/oral response	70.2
Text entry	63.6
Multimedia resources	
Animation/online videos	75.0
Reading materials	
Website	60.0
Information/extended material	78.6
Case diagram	81.8
Interactive/simulation games	100.0

use this tool said it would be more direct to simply ask students to cast their vote by raising their hands. Nonetheless, all teachers said they would be more willing to use it if the voting results could be displayed in the form of charts instead of figures only.

Among the three types of assessment engine tools, selection/oral response (70 %) was most frequently used, followed by text entry (64 %), and making predictions prior to observation (33 %). Selection/oral response was most welcomed by the teachers because it was straightforward, easy to use, and quick; text entry and making predictions prior to observation were less popular among the teachers because they involved more teaching procedures. As stated earlier, the Chinese typing skills of the primary school students were limited; therefore, the teachers would rather let students write than type in such a tight curriculum.

Multimedia resources were more popular than functional tools among the teachers. The percentage of use of the interactive/simulation games reached 100 %. During the interviews, the teachers said that diversified and vivid presentation effects could enhance students' interest in learning. The second most frequently used multimedia resource was case diagrams (82 %), followed closely by news information/extended material (79 %); both were reading materials that students read and studied to gain information about a topic. Animation/online video was also a popular choice of teachers with up to 75 % use. The least frequently used multimedia resource was websites (60 %).

7.3.2.2 Common Ways of Using e-Learning Tools and Resources

Analysis of the recorded lessons showed that the teachers used the various e-learning resources for student learning in three ways: teacher–whole-class participation, group learning, and individual learning. In addition, the teachers used other supplementary classroom activities such as discussion/question and answer, first-hand experience, voting, presentation/sharing, teacher feedback, material reading, and video watching.

Online discussion forum: The teachers first displayed the e-learning resources and explained the learning activities. Some teachers divided the students into small groups to view the e-learning resources and conduct in-class discussions before the students completed the activities on the online discussion forum at home. These teachers would display the results of the online discussion forum and provide feed-back in the next lesson. Other teachers gave worksheets to the students to guide their small-group discussions and would provide feedback after the student presenta-tions. Other teachers led whole-class discussion and provided feedback based on students' responses. The online discussion forum allows asking and discussing to occur among students and between students and teachers. Figure 7.9 shows the common ways of using the online discussion forum.

Online voting: Teachers first displayed the e-learning resources and explained the learning activities. Some teachers moved directly to the voting activity by asking students to raise their hands; others allowed students to have a small-group discus-sion before asking them to cast their votes by either raising their hands or using the online voting system. All the teachers ended the voting activity by inviting students to explain their choices and providing feedback to them. Figure 7.10 shows the common ways of using online voting by the teachers.

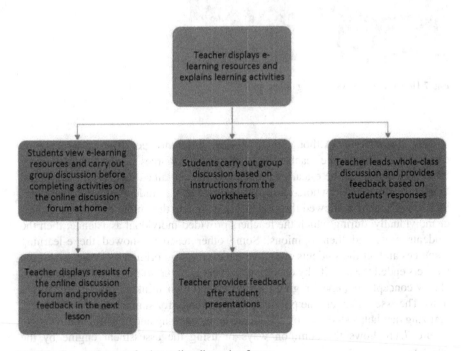

Fig. 7.9 Common ways of using online discussion forum

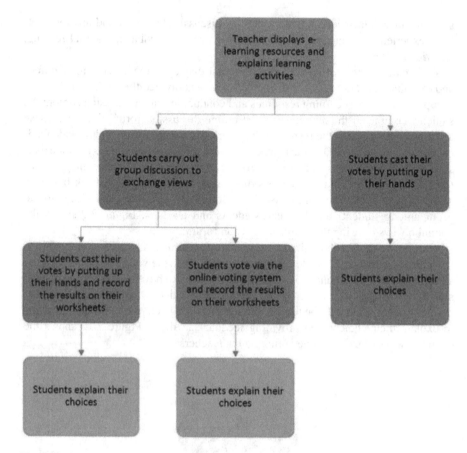

Fig. 7.10 Common ways of using online voting

Assessment engine: Although there are three different types of assessment engine e-learning resources, the teachers used the three resources in quite similar ways. First, they displayed the e-learning resources and explained the learning activities. Some teachers then led whole-class discussion to let the students express their opinions. Some teachers allowed the students to complete the online exercise in groups or individually, during which the teachers provided individual assistance; then the students presented their opinions. Some other teachers showed the e-learning resources and let the students complete the exercise on printed worksheets. All the teachers ended the activity by checking answers with the students and helping them clarify concepts and encouraging them to extend their thinking to daily-life situations. The assessment engine provides an opportunity for summarizing and conceptualizing that helps students to consolidate understanding and construct knowledge. Figure 7.11 shows the common ways of using the assessment engine by the teachers.

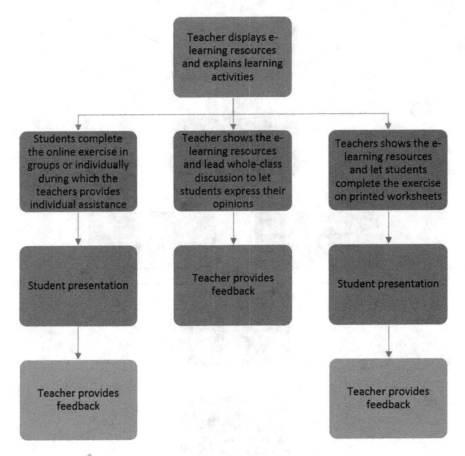

Fig. 7.11 Common ways of using assessment engine

Animation/online videos: Some teachers first displayed the e-learning resources and explained the learning activities; then the students viewed the animation/ online videos with their computers or the teacher's computer; after that, the students completed the exercise on the printed worksheets or conducted discussions. Some other teachers posed questions for students to discuss and think about before viewing the animation/online videos; after viewing the animation/online videos, the teachers led the discussion with the students. To be able to answer the questions on the worksheets or those posed by the teachers, the students had to go through the process of searching and selecting while viewing the animation or online videos. Figure 7.12 shows the common ways of using animation/online videos by the teachers.

Reading materials: The teachers first displayed the e-learning resources, explained the learning activities, and then posed questions as guidance for the students to understand the reading materials. Some other teachers let the students complete the exercise on the printed worksheets. Some teachers let the students complete

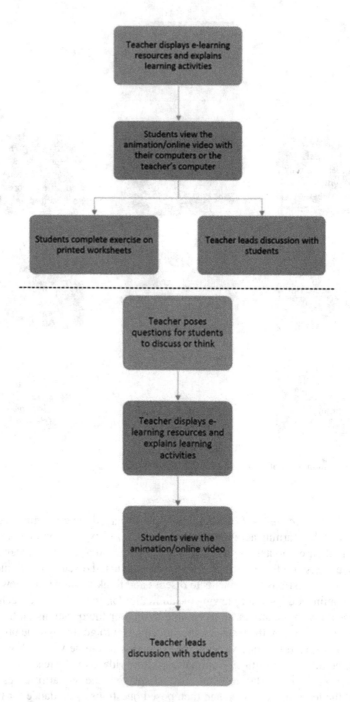

Fig. 7.12 Common ways of using animation/online videos

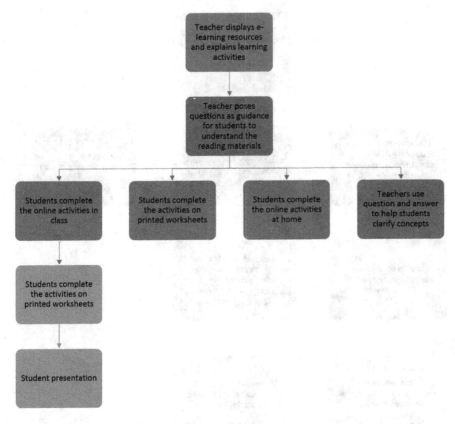

Fig. 7.13 Common ways of using reading materials

the online activity at home, while some simply used questions and answers to help the students clarify concepts. There were also teachers who first let the students complete the online exercise, followed by the exercise on the printed worksheets, and finally make presentations to share their views. Students developed understanding of the topic by searching and selecting relevant information from the reading materials. Figure 7.13 shows the common ways of using the reading materials by the teachers.

Interactive/simulation games: The teachers first displayed the e-learning resources and explained the learning activities. Some teachers let the students complete the game with their computers during which time they provided the students with individual guidance; then the students completed the exercise on the printed worksheets; finally, the students made presentations to share their results and the teachers provided feedback. Some other teachers posed questions for the students to think about before allowing them to play the interactive/simulation games; then they selected some students to use the teacher's computer to play the games while the rest of the class observed; finally, the students completed the exercise on the

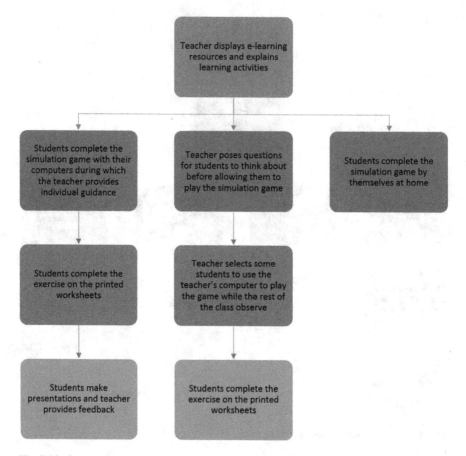

Fig. 7.14 Common ways of using interactive/simulation games

printed worksheets. Some teachers would let the students complete the interactive/ simulation game activity by themselves at home. For games that simulate real experiments (e.g., the ones in closed circuits and the effects of electricity), the students learn through doing and observing as the games serve as tools for information and data collection, allowing students to gather the necessary information and data to answer the inquiry questions. Figure 7.14 shows the common ways of using the interactive/simulation games by the teachers.

None of the 16 teachers interviewed (3 participating teachers were not available for interviews) used technology to address learning differences. Five teachers believed that the learning differences between students were insignificant or that the learning materials or activities were able to match the students' ability so there was no need to use technology or other methods to address learning differences. The remaining 11 teachers said they used other methods to address learning differences, including using different questioning strategies, modifying the worksheets, adopting cooperative learning strategies, and providing more explanations or feedback.

Table 7.3 Paired samples *t*-tests of students' pre- and postlesson tests

Topic	School (*n*)	Pretest *M* (*SD*)	Posttest *M* (*SD*)	*t*	*p**
Digestive system	A (170)	9.4 (1.7)	13.6 (2.2)	−24.68	.001
	B (120)	9.1 (1.7)	12.2 (1.6)	−15.25	.001
Closed circuits	C (30)	11.3 (2.2)	15.8 (1.5)	−11.33	.001
	D (90)	11.4 (2.8)	12.4 (3.0)	−2.96	.004
Effects of electricity	E (120)	9.4 (2.2)	12.0 (1.6)	−9.2	.001
Endangered animals and plants	F (60)	2.2 (1.4)	2.7 (1.6)	−3.4	.001

*p < .01

7.3.2.3 Challenges in Using the e-Learning Resources

All of the teachers reflected that they encountered different problems during the actual implementation of the lessons. The most commonly mentioned problem was the instability and the slow speed of the online learning system, which affected the teaching progress. Other problems related to the online learning platform included the lack of realistic inquiry activities, the system not being optimized for home use due to the lack of instruction, and small font size. In addition, classroom management was a challenge to the teachers because both the teachers and the students were relatively new to teaching and learning with the online learning platform. Therefore, some teachers used collaborative learning to make classroom management easier. In order to help the students stay focused, some teachers asked the students to close the lid when their laptops were not in use.

7.3.3 *Student Learning with the e-Learning Resources*

Paired samples *t*-tests of the pre- and postlesson scores of the students in the six classes suggested that there was significant improvement in student learning after use of the e-learning resources for all the topics (Table 7.3).

7.4 Conclusion

The role the teacher plays in creating and maintaining the course contents varies from a tutor working with materials and instructional design created by others to a "lone ranger" or teacher who creates all of the content (Anderson, 2008). Making reference to the new concept of e-learning that effectively integrates technology, pedagogy, and CK as well as the findings of this study, four suggestions are offered to make e-learning more dynamic in primary classrooms: adjusting the learning content, extending the flexibility of technology use, using diversified activities, and using assistive tools.

7.4.1 Adjusting the Learning Content

Although teachers may not be able to modify the learning content provided by the online learning system, they could use different methods to adjust the content. This would resemble the case in traditional face-to-face teaching where a teacher is asked to teach a syllabus created by someone else. It would be wise for the teacher to determine the coverage of the content provided by the online learning system and to decide what to retain, use, add, or replace (Ko & Rossen, 2010).

For example, for the topic on effects of electricity, the online learning material shows a diagram of an iron nail connected to a closed circuit with a question whether paper clips will be attracted by the iron nail if the circuit is powered. Teachers may add a diagram to the printed worksheets showing a circuit without batteries and ask students whether an iron nail connected to an unpowered circuit can attract the paper clips. This supplementary exercise, combined with the original learning content available on the online learning platform, can test whether the students really understand the magnetic effect of electric currents.

7.4.2 Extending the Flexibility of Technology Use

There is no right way to integrate technology into the classroom; the key is providing students with the most effective learning environment (Johnson & Lamb, 2000/2007). The e-learning resources make use of different technologies to realize the teaching and learning goals; however, the technologies may not be applicable to every school or classroom. Teachers may work on the pedagogical level to utilize different technologies by adding, replacing, or rearranging.

For example, for the topic on digestive system, online voting was designed for students to cast their votes on the activities that are suitable for them to do after a meal. In case the learning environment does not support the use of this technology, teachers can arrange students to show their opinions by raising their hands, a commonly used method in the classroom.

Another example relates to the closed circuits topic; the online learning platform provides two simulation games for students to learn the concept of closed circuits. The first game provides only a few circuits with very limited combinations. The second game offers a greater variety by providing more circuits and allowing the change of parameters so that the students can observe the changes in electric currents and light intensities of different combinations. Based on school-based characteristics (e.g., resources, students' interests and abilities), teachers may let students play the first game, the one with less variety, with real objects (e.g., batteries, wires, and lightbulbs); then when students master the game and grasp the concept, they can move on to the more advanced simulation game provided on the online learning platform, the one that offers greater variety.

7.4.3 Using Diversified Activities

Brown and Voltz (2005) believed that having tasks that are likely to lead students to the desired learning outcomes is key to effective e-learning. Before, during, and after the use of any e-learning resource, teachers can add any activities for providing feedback or getting students to share, discuss, raise questions, or have firsthand experience. These additional activities will enhance the interactions between students and their teachers, peers, and even the e-learning resources and will subsequently enhance the efficacy of the teaching and learning. For example, for the digestive system topic, teachers may let students take some candies to experience how food is digested, and then follow up by asking them to think about in which organ food is first digested before showing the video that introduces the functions of different digestive organs.

7.4.4 Using Assistive Tools

The provision of additional assistive guidance facilitates students to carry out learning activities with the e-learning resources. Printed versions of the online exercise can serve as guides for students throughout the learning process and facilitate the management of classroom teaching and learning. For example, for the endangered animals and plants topic, students are required to enter into the online learning platform what they learned from the websites so that the system can check their answers. The printed worksheets were prepared for students to record the information while browsing the websites and later make use of the answer-checking function of the online learning platform to evaluate their answers written on the worksheets. This matches the website browsing activity and enhances teaching management.

In summary, to realize the concept of TPACK in unique educational contexts, the teachers in this study created online learning environments for students with the existing e-resources available on the online learning platform and facilitated student learning with a variety of pedagogies. This showed that rather than creating online learning environments from scratch, teachers can apply TPACK and use different methods to adjust the learning content, extend the flexibility of technology use, and use diversified activities and assistive tools to suit the needs of the students and to make learning more dynamic and flexible—adapt not adopt in totality.

References

Anderson, T. (2008). Teaching in an online learning content. In T. Anderson (Ed.), *The theory and practice of online learning* (2nd ed., pp. 343–365). Edmonton, AB: AU Press.

Blumenfeld, P. C., Kempler, T. M., & Krajcik, J. S. (2006). Motivation and cognitive engagement in learning environments. In R. K. Sawyer (Ed.), *The Cambridge handbook of the learning sciences* (pp. 475–488). Cambridge, UK: Cambridge University Press.

Brown, A. R., & Voltz, B. D. (2005). Elements of effective e-learning design. *The International Review of Research in Open and Distance Learning, 6*(1). Retrieved from http://www.irrodl.org/index.php/irrodl/article/view/217/300

Curriculum Development Council. (2011). *General studies for primary schools curriculum guide (primary 1 – primary 6)*. Hong Kong, Hong Kong: Author. Retrieved from http://www.edb.gov.hk/attachment/en/curriculum-development/cross-kla-studies/gs-primary/gs_p_guide-eng_300dpi-final%20version.pdf

Doering, A., Veletsianos, G., Scharber, C., & Miller, C. (2009). Using the technological, pedagogical, and content knowledge framework to design online learning environments and professional development. *Journal of Educational Computing Research, 41*(3), 319–346.

Government of the Hong Kong Special Administrative Region. (2011). *The 2011–12 policy address: From strength to strength*. Hong Kong, Hong Kong: Author. Retrieved from http://www.policyaddress.gov.hk/11-12/eng/index.html

Hill, J. R., & Hannafin, M. J. (2001). Teaching and learning in digital environments: The resurgence of resource-based learning. *Educational Technology Research and Development, 49*(3), 37–52.

Johnson, L., & Lamb, A. (2000/2007). *Technology and multiple intelligences*. Retrieved from http://eduscapes.com/tap/topic68.htm

Ko, S., & Rossen, S. (2010). *Teaching online: A practical guide*. New York, NY: Routledge.

Koehler, M. J., & Mishra, P. (2005). Teachers learning technology by design. *Journal of Computing in Teacher Education, 21*(3), 94–102.

Koehler, M. J., Mishra, P., Akcaoglu, M., & Rosenberg, J. (2013). The technological pedagogical content knowledge framework for teachers and teacher educators. In R. Thyagarajan (Ed.), *ICT integrated teacher education: A resource book*. New Delhi, India: CEMCA. Retrieved from http://cemca.org.in/ckfinder/userfiles/files/ICT%20teacher%20education%20Module%201%20Final_May%2020.pdf

Koehler, M. J., Mishra, P., & Yahya, K. (2007). Tracing the development of teacher knowledge in a design seminar: Integrating content, pedagogy, & technology. *Computers and Education, 49*(3), 740–762.

Mishra, P., & Koehler, M. J. (2006). Technological pedagogical content knowledge: A framework for teacher knowledge. *Teachers College Record, 108*, 1017–1054.

Niess, M. L. (2001). Research into practice: A model for integrating technology in preservice science and mathematics content-specific teacher preparation. *School Science and Mathematics, 101*(2), 102–109.

Niess, M. L. (2011). Investigating TPACK: Knowledge growth in teaching with technology. *Journal of Educational Computing Research, 44*(3), 299–317.

Shulman, L. S. (1986). Those who understand: Knowledge growth in teaching. *Educational Researcher, 15*(2), 4–14.

So, W. W. M. (2012). Creating a resource-based e-learning environment for science learning in primary classrooms. *Technology, Pedagogy and Education, 21*(3), 317–335.

Part IV
Epilogue

Chapter 8
The End of the Beginning: An Epilogue

Punya Mishra and Danah Henriksen

This final chapter serves as the epilogue, as both a summary and a synthesis of the chapters in the book. We begin by providing an informal historical overview of the current impact of TPACK as a theoretical framework in terms of the quantifiable reach of the theory as well as the rapidity and breadth of its acceptance. We then provide an overview of each chapter that includes, first, how they are grouped thematically and, then, its core ideas. For Chaps. 2, 3, 4, 5, 6, and 7, we identify and summarize a few key takeaways and points of interest. Following this overview, we identify three crosscutting themes: the importance of the idea of learning by design for the development of TPACK; an emphasis on the evaluation and measurement of TPACK; and, finally, the important role that communities of practice play in TPACK development. We note how learning by design is relevant because several of the studies here involved educators working through the design process (creating software applications, lessons, and other teaching artifacts) to extend it into the arena of TPACK research. Evaluation/measurement is important as well because the work in this book seek to develop rubrics that would allow teacher educators to evaluate different facets of TPACK. Communities of practice were also relevant because, rather than looking at teachers in isolation, the work in this book represents settings that support partnership/teamwork between preservice and in-service teachers (as well as educational researchers, teacher educators, and others). Finally, after considering these points, we offer a note of both positive points and constructive critique regarding this book's potential contributions to the internationalization of TPACK research.

P. Mishra (✉) • D. Henriksen
Department of Counseling, Educational Psychology and Special Education,
Michigan State University, East Lansing, MI, USA
e-mail: punya@msu.edu; henrikse@msu.edu

© Springer Science+Business Media Singapore 2015
Y.-S. Hsu (ed.), *Development of Science Teachers' TPACK*,
DOI 10.1007/978-981-287-441-2_8

Now this is not the end. It is not even the beginning of the end. But it is, perhaps, the end of the beginning.—Winston Churchill (November 10, 1942)

It gives us great pleasure to write the epilogue for this collection of research articles related to technological pedagogical content knowledge (TPACK). The idea of TPACK has truly had a significant impact on the research and practice in educational technology. Speaking personally, it was sometime in 2000 that Matt Koehler and the first author started working together on the learning by design seminars. These seminars, which ended up becoming a book entitled *Faculty Development by Design* (Mishra, Koehler, & Zhao, 2007), were an intervention that attempted to get faculty in higher education to intelligently integrate technology in their teaching. It was while we were conducting research on the process by which faculty working in design teams with graduate students came up with solutions to technological and pedagogical problems of teaching subject matter that the initial idea of TPACK came to us. At that point, it was an inchoate form of understanding—and one that needed further research to elucidate. What I do know is that both Matt and I had a sense that we were closing in on an interesting idea and one that we needed to share with the world. It was around 2004 that we begin writing the article that would finally be published in *Teachers College Record* in 2006.

To say that this article changed our lives is an understatement. The article has over 2,000 citations in Google Scholar. It in turn led to the *Handbook of TPCK* published by Routledge and the American Association of Colleges of Teacher Education (AACTE, Herring, Mishra, & Koehler, 2008). For instance, a quick review of the public Mendeley bibliography connected to the TPACK.org website reveals that there are over 630 publications tagged as being related to TPACK (35 book chapters, 220 conference papers, 15 miscellaneous pieces, and the remainder are journal articles). That is a staggering number of publications—for a topic that was introduced to the research and scholarly community less than a decade ago. In more practical terms, the TPACK framework has been used for faculty development in higher education; it has become an integral part of teacher education and teacher professional development in many countries around the world; and it has been accepted as a guiding framework by a range of educational organizations. As must be clear, the rapidity and breadth of acceptance of the framework have been incredibly gratifying to us. Also gratifying is this opportunity to read all the chapters in this book and to be asked to write an epilogue.

That said, we approach this task with humility; and we do so for two key reasons. First, because though one of the authors of this epilogue is identified as being one of the originators of the framework, we know well that there are many others who have made similar arguments but were not lucky enough to receive the recognition we did (We have in our writing attempted to provide credit to these precursors of the TPACK framework as often as we can.). Second, and as importantly, we understand that the literature on TPACK has grown so quickly that it is nearly impossible for us to keep up with all the work being presented and published. In fact, it can be argued that there are other scholars who are more up to date with the TPACK literature. Given these two facts, it must be understood that this chapter not be seen as a definitive *reading* of this book but rather as one possible review.

8.1   Broad Strokes: Overview of the Outcomes

At the broadest level, this book is concerned with the critical issue of teacher education in developing TPACK. And as readers will have noted, the chapters of the book are organized around three central themes of TPACK development, which include TPACK in Teaching Practices, The Transformative Model of TPACK, and The Integrative Model of TPACK.

The overarching focus of this book—examining ways to improve teacher education for the development of TPACK—is relevant and essential to our global and technology-driven society. By improving the way that current and future teachers teach with technology, the field of education ensures that we will meet the needs of twenty-first-century students. Building on the potential of technology offers us a way to enrich and expand learning opportunities and to expand the types of experiences that teachers and learners can have in the classroom.

One of the critical contemporary issues in teacher education has involved how to better support preservice and in-service teachers in the way that they teach with technology. Mishra and Koehler (2006) suggested that this could be well addressed through developing TPACK with the engagement of instructional frameworks, proper assessments of knowledge and practices, and teaching practices for specific learning and teaching contexts. The pre- and in-service focus on educational technology in the chapters of this book highlights an area of teaching and learning that is at the crux of modern education globally. The different frameworks and approaches applied by these authors, along with the different aspects of TPACK they investigated, offer some valuable insights for teacher education and professional development. They are significant as a first step toward a more research-based and informed look at how TPACK is operating in different aspects of teacher learning. Several interesting strands of research arise as we look across the chapters.

8.1.1   TPACK in Teaching Practices

Chapters 2 and 3 are focused on understanding the ways that TPACK is instantiated in practice. Chapter 2 highlights the fact that there has been much research done to consider and study the models and variations of TPACK for different contexts (e.g., TPACK-deep, TPACK-W). As we see it, there is a research gap in which there has been a lack of work examining working models of TPACK within more subject-specific contexts, such as science, mathematics, etc. This is an interesting gap, particularly when we consider the fact that TPACK itself is so tied to content and the way that content explicitly alters teaching practices and uses of technology. It stands to reason that more diversity within models of TPACK could be useful in subject-/content-specific approaches, and this was a core aspect of Chap. 2. A two-strand panel of researchers and expert teachers helped to generate and validate a TPACK-practical (TPACK-P) framework. The knowledge of learners, knowledge of classroom instruction, and knowledge of curriculum design components that they describe not only maps on to existing aspects of TPACK but also considers some subject-specific issues. For example in teaching science content, diversity of representations is particularly meaningful and holds unique considerations for technology. The possibility for a more detailed set of subject-specific models of TPACK is a fascinating and useful approach for adding to the existing body of more generalized TPACK work; we concur with the authors that more work is needed in this area.

In Chap. 3, the authors studied novice and experienced science teachers to better understand their TPACK-P knowledge. They did this via interviews with 40 science teachers to reveal their TPACK-P (along the lines of assessment, planning and designing, and teaching practice). The coding schema they developed is interesting in that it provides the field of TPACK research with three categories of teacher knowledge: infusive application, transition, and plan and design emphasis. These three categories hold possibilities for understanding different levels of teacher fluidity with TPACK, from the more infusive (expert) group to the transition group, and to the plan/design group (who seemed more comfortable with lesson planning and preparation of technology-driven lessons than the actual implementation). This analytical breakdown of different levels of TPACK understanding is significant in that it provides support to teachers at different places in the process of knowing and implementing technology approaches in their teaching. As the authors suggest, the patterns shown in this chapter can become a guiding framework for the development of instruments that evaluate teachers' competence in using classroom technologies. More importantly, it gives us a way to see what they do well and where they struggle. In this sense, it is a useful diagnostic approach to giving teachers (and teacher educators) a look at where they are, and where they can go, when it comes to teaching with technology.

8.1.2   The Transformative Model of TPACK

Chapter 4 puts a focus on research that seeks a deeper understanding of how TPACK is evaluated in science teaching. Specifically, the authors created and tested rubrics to evaluate preservice teachers' TPACK-P; and these were developed according to the proficiency levels and features previously identified about in-service teachers. They collected lesson plans and microteaching video clips of preservice teachers working on physics curriculum and instruction design. Interestingly, results revealed that these preservice teachers' performances on lesson planning and microteaching were similar within one level of proficiency. However, their performance on teaching with technology was comparatively better in curriculum design and enactment than on assessment. In other words, new and future teachers have an easier time in the planning/enacting of technology lessons than with assessment.

In Chap. 5, the authors explore a teacher community consisting of a teacher educator, four experienced physics teachers, and 11 preservice teachers who collaborated with each other on developing simulation-based physics learning modules. With experienced teachers designing software applications (Apps) or learning modules, the preservice teachers played the role of not only users who implemented the Apps but also testers/evaluators of the Apps. This study presented an interesting case of learning-through-design work for technology implementation and knowledge; it did so in a model that worked for different proficiency levels. The more experienced App designers refined their TPACK-P while producing and reflecting on the artifacts. And the testing and evaluation process gave the preservice teachers an opportunity to experience variables and visualize the phenomena and how it operates in teaching and learning settings. More importantly, this chapter reflects the way that communities of practice can be invaluable in teaching with technology situations. The novice teachers were able to learn from and with the expert teachers and vice versa; and the design-centered approach made the task valuable to teachers at all proficiency levels, giving them a chance to grow their TPACK in practical ways.

8.1.3 The Integrative Model of TPACK

Chapter 6 bases its work on the theoretical framework of cognitive apprenticeship. The authors apply the MAGDAIRE model (modeled analysis, guided development, articulated implementation, and reflected evaluation) to help preservice teachers become more sensitive to the interplay between the elements of TPACK. This model seemed to be a useful framework for allowing preservice teachers to consider how technology connects to their teaching practice based on a set of variables. The authors found that the preservice teachers they worked with moved toward a more connected look at the ways that technology intertwines with teaching school subject matters. We liken their model to an effective mingling of the cognitive apprenticeship learning theory with a detailed learning-by-design framework. In this, it provides an approach to improve preservice teachers' TPACK that is supportive, collaborative, and systematic (tapping the knowledge of expert teachers for novices, within a guided framework).

In Chap. 7, the authors had teachers utilize e-learning resources of four science topics in the primary curriculum in order to observe and learn from the ways in which they applied this technology. The results from the 19 teachers invited to use these e-learning resources in their classrooms showed some specific understandings of how technology supports teaching and learning. The range of findings seemed valuable for presenting a look at how teachers use technology in a very broad context. Though this work was done in Hong Kong, many of the issues that arose have applicability in many other countries and settings (certainly in the USA). Some of the teachers' initial concerns about technology implementation included the following: worrying whether implementing such activities would affect the teaching schedule and, consequently, students' examination results; the adequacy of different technology equipment; the importance of the teaching materials matching the content of the textbooks or school-based curriculum. These are similar to broader issues faced by all educators who seek to intelligently incorporate technology in their teaching. The authors also note a need for flexibility, in that teachers might need to modify technology resources or content to fit the resources. But most importantly, they derive the conclusions that (a) there is no right way to integrate technology into the classroom and (b) applicability is highly variable based on the classroom and the context.

8.2 Thematic Issues

8.2.1 Learning by Design

One important guiding theme that we found interesting and important throughout several of these studies was that a type of learning-by-design framework was sometimes implemented to help teachers learn and expand their TPACK. Learning by design is an approach in which learners construct their knowledge through the process of creating something (Kafai, 1995)—quite literally, learning by going through the design process (Shaltry, Henriksen, Wu, & Dickson, 2013). In recent years, this approach has increased in significance in learning/technology research, especially in relation to constructionist frameworks (Peppler & Kafai, 2010; Wiggins & McTighe, 2005). Several of the studies involved instances of educators working through the design process (creating applications, lessons, and other teaching artifacts), which extends the learning-by-design approach into the arena of TPACK research. Some of this design work invites preservice and in-service teachers to work together, which is an approach that fits well with TPACK understandings and with the dynamic and social interplay of the factors that make it up. Mishra and Koehler (2006) suggested that learning by design is a foundational way of thinking and learning, in building a mindset for TPACK. We highlight this point because the development of TPACK is a relatively sophisticated type of expertise that takes educators time and efforts across years to develop. But in a fundamentally important way, the learning in practice that happens in design-based approaches is an excellent way to set the stage for this among new and future teachers. In the case of practicing teachers, it is an approach to honing their craft and taking their TPACK to the next level, using the skills of a designer.

8.2.2 Evaluation and Measurement

Evaluations of teaching often happen instinctively in the classroom, and they can be a relatively subjective area of teacher education. It is an innately subjective and human activity to observe and make judgments about approaches to teaching and methods of interacting with students, ideas, and technology. It is important, however, that we go beyond mere subjectivity in evaluating teachers, particularly in a realm of teaching as relatively recent as TPACK and digital classroom technologies.

As Lord William Thomson Kelvin once said, not being able to measure what it is that we are speaking of is a "meager and unsatisfactory" kind of knowledge (as cited in Mishra, Henriksen, & Deep-Play Research Group, 2013, p. 11). Toward this purpose, we applaud the efforts of the work in this book aimed at developing rubrics that would allow teacher educators to evaluate different facets of TPACK. The studies in the book are relatively exploratory in entering new territory of educational evaluation. However, the efforts are significant in that they contribute not only through providing some original and early gauges of TPACK in preservice teacher education, but may also be useful for in-service teachers to know their level of proficiency. In fact, the research-based methods for such rubric development constitute a valuable thing as well, for providing the foundations for others to develop new TPACK rubrics in context. It is only through understanding where we are at that we are able to move forward; by offering such measures to teachers, we can help them in their TPACK growth.

8.2.3 Communities of Practice

The work in this book represents substantive TPACK research and findings that were frequently derived through collaboration, communication among teachers, and communities of practice. We were interested and encouraged to note that several of these studies put teachers in a position of learning and developing their TPACK together. Rather than looking at teachers in isolation, the work in this book represents settings with supports and partnership/teamwork between preservice and in-service teachers (as well as educational researchers, teacher educators, and others). Lave and Wenger (1991) showed how communities of practice (e.g., groups of teachers) offer opportunities for learning through informal apprenticeship models. The role of preservice teachers in several of these chapters maps nicely onto this view of learning and fits well with the way that teachers actually operate and learn to teach in the real world of classrooms. This situates the research in a collaborative learning framework and the best possible situation for authentic approaches to TPACK development.

Brown, Collins, and Duguid (1989) described authentic activities as "the ordinary practices of the culture." (p. 34). We note that learning through collaboration is clearly ordinary/authentic practice for teachers. Often times, such situative learning happens during an internship or another field experience. However, the opportunities demonstrated in this research present new avenues for building TPACK through discussion, collaboration, and/or design practices among new and experienced teachers (Shaltry et al., 2013). As Granger, Morbey, Lotherington, Owston, and Wideman (2002) put it, "Like effective leadership, the importance of collaboration cannot be overestimated: teachers need each other—for team teaching and planning, technical problem solving assistance and learning" (p. 486); we think that this translates clearly onto the TPACK research settings in this body of work.

8.3   A Positive Note … and a Point of Critique

For historical, and other contingent, reasons most educational research (and educational technology research) has generally happened in North America, specifically the USA. This is true of TPACK-related research as well. Although exact figures are difficult to come by, a recent review by Chai, Koh, and Tsai (2013) indicated that approximately 65 % of the studies selected were conducted in North America. Europe and the Asia-Pacific region were evenly matched at around 17 %.

Given the forces of globalization and the spread of technology, it is clear that there needs to be a better, and fairer, distribution of research. This is particularly true when we think of the important role the *dotted circle* (Chap. 1) of context plays in the TPACK diagram. Therefore, this book focusing on outside the USA is a helpful corrective to the inordinate emphasis on US-based contexts of educational technology research. Through this internationalization of research and work that examines TPACK in a more varied, broad, and global context, we get a better sense of how the framework plays out in practice from different perspectives. We think that this book is an important step toward that goal and that there needs to be more work of this sort that looks at TPACK in international contexts. It is essential that international educational technology research (like the studies in this book) further our understanding of TPACK in a global way, rather than a narrower, strictly American consideration of the framework.

An important question then becomes: *How do international contexts differ?* We have some understanding of contexts in the USA already, but a broader look at different TPACK contexts is useful for the future and worthwhile to examine. Going beyond western educational settings, it is important to connect these ideas globally and learn through comparisons and contrasts. The work in this book speaks to the value of a framework such as TPACK because, without a framing structure, individual studies would be difficult to connect to a larger picture in education. The framework brings these ideas together and gives us something to connect and compare/contrast between different settings and instantiations of TPACK. Thereby, we applaud the authors and editor of this book for providing research in another set of contexts that adds substantially to the *big picture* of TPACK and educational technology.

That said, we would be remiss if we did not offer observations or critique that could add even more to the body of work going forward. So, one criticism of the book could be the lack of contextual information provided in each chapter. Providing any broader contextual information about educational technology or e-learning could be a useful way to lay the groundwork. For example, the size of the e-learning markets in ten Asian countries was the central focus of a recent report (Bashar & Khan, 2007). Korea, China, and Singapore were the three largest markets in 2002; and Taiwan was ranked sixth. Though Taiwan has a comparatively smaller market, the government there supported efforts to build the e-learning related infrastructure (e.g., educational technology availability in classrooms, Science Park for technology advancement), curriculum reform, and friendly policies for e-learning industries

(Qi, 2005). This type of information helps to set the stage for helping international readers understand the broader context.

While providing this type of national or market-based information is useful, it is also essential to include more localized contexts, such as classroom size, teacher professional development, and so on. We would argue that truly understanding TPACK (and its instantiations in specific classrooms) may require going even deeper. For instance, what are the cultural parameters within which teachers and classrooms function? What is the role of the teacher in the culture of the classroom? What is the culture overall? And how do these views and approaches to teaching relate to the use of TPACK and educational technology? These are just a few possible issues or questions that could be interesting to consider, or to include some thoughts on, as we seek to expand the borders of TPACK research.

All of this attention to context is important in order to avoid perpetuating the myth that educational contexts do not matter—a myth that has too long been a part of educational research. It is always good to deepen the understanding of context with rich, clarifying detail. Educational technology is constructed as much by wires and devices as it by social constraints and policies and politics. It is imperative that we develop a better understanding of these contextual matters. In this respect, this book is an excellent and positive step forward; and it allows us to see even more possibilities for the future of these lines of rich global research.

The TPACK framework has spread its wings and established itself in the arena of educational technology since its first public presentation in 2006. That said, this book and the chapters within it indicate that there is still much interesting work being done today and more that needs to be done in the future. In that sense, we are nowhere near the end of the journey but we are, possibly (as Churchill said in a somewhat different context), at the end of the beginning.

References

American Association of Colleges of Teacher Education Committee on Innovation and Technology, Herry, M., Mishra, P., & Koehle, M. J. (Eds.). (2008). *Handbook of technological pedagogical content knowledge (TPCK) for educators*. New York, NY: Routledge.

Bashar, M. I., & Khan, H. (2007). *E-Learning in Singapore: A brief assessment* (U21Global Working Paper Series, No. 003/2007). Available at SSRN: http://ssrn.com/abstract=1601611 or http://dx.doi.org/10.2139/ssrn.1601611

Brown, J. S., Collins, A., & Duguid, P. (1989). Situated cognition and the culture of learning. *Educational Researcher, 18*(1), 32–42.

Chai, C. S., Koh, J. H. L., & Tsai, C.-C. (2013). A review of technological pedagogical content knowledge. *Educational Technology & Society, 16*(2), 31–51. Retrieved from http://www.ifets.info/abstract.php?art_id=1349

Granger, C. A., Morbey, M. L., Lotherington, H., Owston, R. D., & Wideman, H. H. (2002). Factors contributing to teachers' successful implementation of IT. *Journal of Computer Assisted Learning, 18*(4), 480–488.

Kafai, Y. B. (1995). *Minds in play: Computer game design as a context for children's learning*. Hillsdale, NJ: Lawrence Erlbaum.

Lave, J., & Wenger, E. (1991). *Situated learning: Legitimate peripheral participation*. Cambridge, MA: Cambridge University Press.

Mishra, P., & Koehler, M. J. (2006). Technological pedagogical content knowledge: A framework for teacher knowledge. *Teachers College Record, 108*(6), 1017–1054.

Mishra, P., Henriksen, D., & Deep-Play Research Group. (2013). A NEW approach to defining and measuring creativity. *Tech Trends, 57*(5), 5–13.

Mishra, P., Koehler, M. J., & Zhao, Y. (Eds.). (2007). *Faculty development by design: Integrating technology in higher education*. Charlotte, NC: Information Age.

Peppler, K., & Kafai, Y. B. (2010). Gaming fluencies: Pathways into a participatory culture in a community design studio. *International Journal of Learning and Media, 1*(4), 1–14.

Qi, J. J. (2005). The gap between e-learning availability and e-learning industry development in Taiwan. In A. A. Carr-Chellman (Ed.), *Global perspectives on e-learning: Rhetoric and reality* (pp. 35–51). Thousand Oaks, CA: Sage.

Shaltry, C., Henriksen, D., Wu, M. L., & Dickson, P. (2013). Teaching pre-service teachers to integrate technology: Situated learning with online portfolios, classroom websites and Facebook. *TechTrends, 57*(3), 20–25.

Wiggins, G., & McTighe, J. (2005). *Understanding by design* (2nd ed.). Alexandria, VA: Association for Supervision and Curriculum Development.

Author Index

Subject Index

© Springer Science+Business Media Singapore 2015
Y.-S. Hsu (ed.), *Development of Science Teachers' TPACK*,
DOI 10.1007/978-981-287-441-2

Printed in the United States
By Bookmasters